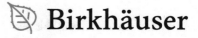

Trends in Mathematics

Research Perspectives CRM Barcelona

Volume 9

Series editor

Enric Ventura

Since 1984 the Centre de Recerca Matemàtica (CRM) has been organizing scientific events such as conferences or workshops which span a wide range of cutting-edge topics in mathematics and present outstanding new results. In the fall of 2012, the CRM decided to publish extended conference abstracts originating from scientific events hosted at the center. The aim of this initiative is to quickly communicate new achievements, contribute to a fluent update of the state of the art, and enhance the scientific benefit of the CRM meetings. The extended abstracts are published in the subseries Research Perspectives CRM Barcelona within the Trends in Mathematics series. Volumes in the subseries will include a collection of revised written versions of the communications, grouped by events.

More information about this series at http://www.springer.com/series/13332

Maria Alberich-Carramiñana
Carlos Galindo · Alex Küronya
Joaquim Roé
Editors

Extended Abstracts February 2016

Positivity and Valuations

 Birkhäuser

Editors
Maria Alberich-Carramiñana
Departament de Matemàtiques
Universitat Politècnica de Catalunya
Barcelona, Spain

Alex Küronya
Institut für Mathematik
Goethe-Universität Frankfurt
Frankfurt, Germany

Carlos Galindo
Departament de Matemàtiques
University Jaume I
Castellón, Spain

Joaquim Roé
Departament de Matemàtiques
Universitat Autònoma de Barcelona
Bellaterra, Barcelona, Spain

ISSN 2297-0215   ISSN 2297-024X   (electronic)
Trends in Mathematics
ISSN 2509-7407   ISSN 2509-7415   (electronic)
Research Perspectives CRM Barcelona
ISBN 978-3-030-00026-4   ISBN 978-3-030-00027-1   (eBook)
https://doi.org/10.1007/978-3-030-00027-1

Library of Congress Control Number: 2018955169

Mathematics Subject Classification (2010): 13A18, 13F30, 14T05, 14C20, 14E15, 14E30

This book is published under the imprint Birkhäuser, www.birkhauser-science.com, by the registered company Springer Nature Switzerland AG
The registered company address is: Gewerbestrasse 11, 6330 Cham, Switzerland

Preface

This volume collects Extended Conference Abstracts originated at the workshop "Positivity and Valuations", held at the Centre de Recerca Matemàtica in February 2016. This workshop brought together a variety of researchers, some of them experts on valuations, others interested in their use in the study of positivity in Algebraic Geometry.

Valuation theory was initiated by Kürschák for treating the theory of p-adic fields more than a century ago; it has been flourishing ever since, with deep connections to algebraic number theory, algebraic geometry and the theory of ordered fields. Much of algebraic number theory can be better understood by using valuation theoretic methods, and the same principle applies to the resolution of singularities or the structure of singularities as realized by Zariski and Abhyankar.

Having been dormant for a while in algebraic geometry, there was a recent surge of interest as a tool to attack the exciting open problem of resolution in positive characteristic and to analyze the structure of singularities. As important examples for the expansion of valuation theory, analogues of the Riemann–Zariski valuation spaces have been found to be the right framework for questions of intersection theory in the algebraic geometry and in the analysis of singularities of complex plurisubharmonic functions.

In a different direction, the relation between Berkovich geometry, tropical geometry, and valuation spaces, on the one hand, and the geometry of arc spaces and valuation spaces, on the other, has begun to deepen and clarify. It has to be pointed out that areas listed above account for a significant amount of contemporary research in algebraic and arithmetic geometry.

Considering the connections listed above, it is by no means surprising that valuation theory became relevant for the positivity questions on projective varieties and specifically for the classification problem and the minimal model program (MMP). Already in Lazarsfeld's landmark book on positivity, the connection of valuations with graded linear series and their relevance for open problems on the rationality of asymptotic invariants became clear. The amount of evidence for this relationship grew significantly with the more recent works of Boucksom, Ein, Küronya, Lazarsfeld, Lozovanu, Mustață, and Smith, especially after the introduction and systematic study of Newton–Okounkov bodies. The connection appears to be particularly strong in the case of local positivity of line bundles, and we expect several breakthroughs in this area in the coming years.

Large tracts of birational geometry can be phrased as a study of certain rank one valuations on the field of rational functions, so the significance of valuation theory for the minimal model program is apparent. As shown recently by Galindo and Monserrat, valuation theory and the minimal model program have close ties through the finite generation of Cox rings as well. The connection has its roots in dimension two, since the finite generation of Cox rings can sometimes be tested in terms of positivity of divisors determined by certain valuations. Note that, for varieties with finitely generated Cox rings, the MMP can be carried out for any divisor because the required flips and contractions exist and every sequence terminates.

In the workshop, all these threads of valuation theory in algebraic geometry were present, a group of researchers with different backgrounds, working on valuation theory or interested in the use of valuations for the study of projective algebraic varieties joined, in an effort to exchange views and foster interaction between the different points of view. The main focus was the relationship between valuations and positivity properties of line bundles.

The event consisted of ten talks by distinguished actors of the recent progress outlined above. These talks took place on the first day of the workshop and the remaining morning sessions, and they served to stimulate further discussion. The afternoons were devoted to performing research in working groups on topics in the area of the workshop chosen by the participants at the beginning of the venue. The methodology had already been tested on several occasions (2015 Padova, 2014 Oberwolfach, 2013 Warsaw, 2012 Mainz, 2010 Oberwolfach) with excellent results both on a personal level and as far as mathematical research goes, resulting in new collaborations and results during the week of the workshop and later.

List of Talks

- Félix Delgado (Universidad de Valladolid), *Poincaré series and generating sequences for plane valuations.*
- Victor Lozovanu (Université de Caen Normandie), *From convex geometry of certain valuations to positivity aspects in algebraic geometry.*
- Catriona MacLean (Institut Fourier), *Functions on Newton–Okounkov bodies associated to valuations and some applications.*
- Francisco Monserrat (Universitat Politècnica de València), *The cone of curves and the Cox ring of rational surfaces given by divisorial valuations.*
- Tomasz Szemberg (Pedagogical University of Cracow), *Very general monomial valuations of $\mathbb{P}^2$ and a Nagata type conjecture.*
- Bernard Teissier (Institut de Mathématiques de Jussieu), *Valuations on Noetherian local domains and their associated graded rings.*

- Michael Temkin (Einstein Institute of Mathematics), *Tame distillation and applications to desingularization.*
- Amaury Thuillier (École Normale Supérieure de Lyon), *Piecewise monomial skeleta in Berkovich geometry.*
- Willem Veys (Katholieke Universiteit Leuven), *Semigroup and Poincaré series for a finite set of divisorial valuations.*
- Annette Werner (Goethe Universität Frankfurt), *Non-Archimedean and tropical geometry.*

Additionally, nine young participants presented posters about their recent research:

- Hans Baumers (Katholieke Universiteit Leuven), *Computing jumping numbers in higher dimensions.*
- Guillem Blanco (Universitat Politècnica de Catalunya), *Computing multiple ideals in smooth surfaces.*
- Harold Blum (University of Michigan), *On divisorial valuations computing minimal log discrepancies and log canonical thresholds.*
- Grzegorz Malara (Pedagogical University of Cracow), *On the containment hierarchy for simplicial ideals.*
- Julio Moyano (Universitat Jaume I, Castelló), *The universal zeta function for curve singularities and its relation with global zeta functions.*
- Matthias Nickel (Goethe Universität Frankfurt am Main), *Algebraic volumes of divisors.*
- Jusztina Szpond (Pedagogical University of Cracow), *On Hirzebruch type inequalities and applications.*
- Alejandro Soto (Goethe Universität Frankfurt am Main), *Completion of normal toric schemes over valuation rings of rank one.*
- Laura Tozzo (Technische Universität Kaiserslautern), *Duality on value semigroups.*

Among the full set of seven proposed subjects for working groups, participants distributed themselves into five groups, whose themes were named *Cones on Zariski–Riemann spaces, Local numerical equivalence, Picard number and Newton–Okounkov bodies, Finite generation of valuative semigroups* and *non-Archimedean analytification and Newton–Okounkov bodies.* All groups reported a certain amount of progress on the questions considered, and in most cases the germ of a research paper was started.

In this volume, in addition to extended abstracts of talks and posters presented during the workshop, brief reports on the outcome of working groups are also included. We are very happy to attest that the atmosphere created by the participants of the workshop was very open and friendly, and led to effective collaboration, as can be seen in these group reports.

Bellaterra, Spain
2016

Barcelona, Spain   Maria Alberich-Carramiñana
Castellón, Spain   Carlos Galindo
Frankfurt, Germany   Alex Küronya
Barcelona, Spain   Joaquim Roé

Contents

Newton–Okounkov Bodies of Exceptional Curve Plane Valuations Non-positive at Infinity

Carlos Galindo, Francisco Monserrat, Julio José Moyano-Fernández and Matthias Nickel

Abstract In this note we announce a result determining the Newton–Okounkov bodies of the line bundle $\mathcal{O}_{\mathbb{P}^2}(1)$ with respect to *exceptional curve plane valuations non-positive at infinity*.

1 Introduction

Newton–Okounkov bodies were introduced by Okounkov [19–21] and independently developed in greater generality by Lazarsfeld–Mustaţă [18], on the one hand, and by Kaveh–Khovanskii [13], on the other.

The key idea is to associate a convex body to a big divisor on a smooth irreducible normal projective variety X, with respect to a specific flag of subvarieties of X, via the corresponding valuation on the function field of X. This turns out to be a good

The authors wish to thank J. Roé and A. Küronya for stimulating their interest in Newton–Okounkov bodies as well as for their helpful comments and for pointing out a more customary name for our valuations. The first three authors were partially supported by the Spanish Government Ministerio de Economía y Competitividad (MINECO), grants MTM2012-36917-C03-03 and MTM2015-65764-C3-2-P, as well as by Universitat Jaume I, grant P1-1B2015-02.

C. Galindo (✉) · J. J. Moyano-Fernández
Departament de Matemàtiques, Institut Universitari de Matemàtiques i Aplicacions
de Castelló, Universitat Jaume I, Campus de Riu Sec, 12071 Castellón de la Plana, Spain
e-mail: galindo@uji.es

J. J. Moyano-Fernández
e-mail: moyano@uji.es

F. Monserrat
Institut Universitari de Matemàtica Pura i Aplicada, Universitat Politècnica de València,
Camino de Vera s/n, 46022 València, Spain
e-mail: framonde@mat.upv.es

M. Nickel
FB Informatik und Mathematik, Goethe-Universität Frankfurt, 60054 Frankfurt am Main,
Germany
e-mail: nickel@math.uni-frankfurt.de

© Springer Nature Switzerland AG 2018
M. Alberich-Carramiñana et al. (eds.), *Extended Abstracts February 2016*,
Trends in Mathematics 9, https://doi.org/10.1007/978-3-030-00027-1_1

1

way to relate the convex geometry of that object with positivity aspects on the side of the algebraic geometry. More specifically, Newton–Okounkov bodies seem to be suitable to explain, from their convex structure, the asymptotic behavior of the linear systems given by the divisor and the valuation, as well as the structure of the Mori cone of X and positivity properties of divisors on X; see [2, 14–17].

The computation of Newton–Okounkov bodies is a very hard task and, sometimes, their behaviour is unexpected; see Küronya–Lozovanu [14]. The case when the underlying variety is a surface is also very hard but there exist some known results which can help. We know that they are polygons with rational slopes and can be computed from Zariski decompositions of divisors.

Very recently, Ciliberto–Farnik–Küronya–Lozovanu–Roé–Shramov [4] studied the Newton–Okounkov bodies with respect to exceptional curve plane valuations ν, defined by divisorial valuations ν' with only one Puiseux exponent, centered at a point p in $\mathbb{P}^2 := \mathbb{P}^2_{\mathbb{C}}$, where $\mathbb{C}$ stands for the complex numbers. These valuations have both rank and rational rank equal to 2 and their transcendence degree equals zero. It is proved in [4] that the Newton–Okounkov bodies of the line bundle $\mathcal{O}_{\mathbb{P}^2}(1)$, with respect to exceptional curve plane valuations, are triangles or quadrilaterals, where the vertices are given by the defining Puiseux exponent β', an asymptotic multiplicity $\hat{\mu}$ corresponding with ν' and a value in the segment $[0, \hat{\mu}/\beta']$. The asymptotic multiplicity $\hat{\mu}$ can also be used to formulate a generalization of Nagata's conjecture [6]; see also [11]. The exact value of $\hat{\mu}$ is only known in some cases, including when $\beta' < 7 + 1/9$.

In this note we only announce a result which determines the Newton–Okounkov bodies of the previously mentioned line bundle with respect to any *exceptional curve plane valuation non-positive at infinity*. The proof and additional details will be published in a forthcoming paper.

Exceptional curve plane valuations non-positive at infinity are a large class of exceptional curve plane valuations, that can have any number of Puiseux exponents and are defined by flags $X \supset E \supset \{q\}$, where E is the last exceptional divisor obtained after a simple finite sequence of point blowing-ups starting at $\mathbb{P}^2$, and defines a plane divisorial valuation ν_E which is non-positive at infinity, cf. Galindo–Monserrat [10]. The valuations ν_E are centered at infinity (see Favre–Jonsson [9]) and present a behavior close to that of plane curves with only one place at infinity (see Abhyankar–Moh [1], and Campillo–Piltant–Reguera [3]).

We finish this introduction being more specific and saying that all the mentioned Newton–Okounkov bodies are triangles and we will give their vertices explicitly. Moreover, the anticanonical Iitaka dimension of infinitely many of the considered surfaces X is $-\infty$ and, in addition, their Picard numbers are arbitrarily large.

Recall that the number of vertices of the Newton–Okounkov body defined by a flag and a big divisor on a surface X is bounded by $2\rho + 2$, ρ being the Picard number of X [14], but the above mentioned results in [4] suggest that the bound could be applied even if we consider the flag on a projective model dominating X (and the Newton–Okounkov body associated to the pull-back of a big divisor on X). Our result can be regarded as new evidence supporting this conjecture.

The results presented in this short note were obtained during a visit of the fourth author to the University Jaume I. Previous studies and a large number of computations were done during the workshop *Positivity and valuations* held on February 2016 at the CRM in Barcelona.

2 The General Setting

Let X be a smooth projective variety of dimension n over $\mathbb{C}$. We will write $K(X)$ for the function field of X. Let us fix a flag of subvarieties

$$Y_\bullet := \{X = Y_0 \supset Y_1 \supset Y_2 \supset \cdots \supset Y_n = \{q\}\}$$

such that each $Y_i \subset X$ is irreducible, of codimension i and smooth at q. The point $q \in X$ is called the *center* of the flag.

One may associate to the flag $Y_\bullet$ a discrete valuation of rank n as follows. First, let $g_i = 0$ be the equation of Y_i in Y_{i-1} in a Zariski open set containing q, which is possible since Y_i has codimension i. Then, for $f \in K(X)$ we define

$$\nu_1(f) := \mathrm{ord}_{Y_1}(f), \quad f_1 = \left.\frac{f}{g_1^{\nu_1(f)}}\right|_{Y_1}$$

and, for $2 \le i \le n$, $\nu_i(f) := \mathrm{ord}_{Y_i}(f_{i-1})$, where $f_i = f/g_{i-1}^{\nu_{i-1}(f_{i-1})}|_{Y_i}$. Then, the map $\nu_{Y_\bullet} : K(X) \setminus \{0\} \to \mathbb{Z}^n_{\mathrm{lex}}$ defined by the sequence of maps ν_i, $1 \le i \le n$, as $\nu_{Y_\bullet} := (\nu_1, \ldots, \nu_n)$ is a rank n discrete valuation and any maximal rank valuation comes from a flag [4, Th. 2.9]. Given a flag $Y_\bullet$ and a Cartier divisor D on X, the following subset of $\mathbb{R}^n_+$:

$$\Delta_{Y_\bullet}(D) := \overline{\bigcup_{m \ge 1} \left\{ \frac{\nu_{Y_\bullet}(f)}{m} \mid f \in H^0(X, mD) \setminus \{0\} \right\}},$$

where $\overline{\{\,\cdot\,\}}$ stands for the closed convex hull, is called to be the *Newton–Okounkov body* of D with respect to $Y_\bullet$.

Newton–Okounkov bodies are convex bodies such that

$$\mathrm{vol}_X(D) = n!\,\mathrm{vol}_{\mathbb{R}^n}\left(\Delta_{Y_\bullet}(D)\right),$$

where $\mathrm{vol}_{\mathbb{R}^n}$ means Euclidean volume and

$$\mathrm{vol}_X(D) := \lim_{m \to \infty} \frac{h^0(X, mD)}{m^n/n!}.$$

Moreover, given $D \neq D'$ two big divisors on X, they are numerically equivalent if and only if the associated Newton–Okounkov bodies coincide for all admissible flags on X; see [12]. Furthermore, in the case of surfaces, D and D' are numerically equivalent (up to negative components in the Zariski decomposition that do not go through q) if and only if the associated Newton–Okounkov bodies coincide for all flags centered at q, cf. Roé [22].

3 Exceptional Curve Plane Valuations and Newton–Okounkov Bodies

In this section, we introduce the family of flags for which we are interested in computing Newton–Okounkov bodies. Let $\mathbb{P}^2$ be the complex projective plane, and p any point in $\mathbb{P}^2$. Let R be the local ring of $\mathbb{P}^2$ at p, and write F for the field of fractions of R. Valuations ν of F centered at R are in one-to-one correspondence with simple sequences of point blowing-ups whose first center is p [23, p. 121]:

$$\eta: \quad \cdots \longrightarrow X_n \longrightarrow X_{n-1} \longrightarrow \cdots \longrightarrow X_1 \longrightarrow X_0 = \mathbb{P}^2.$$

The *cluster of centers* of η will be denoted by $\mathcal{C} = \{p = p_1, p_2, \ldots\}$ and we say that a point p_i is proximate to p_j, $i > j$, written $p_i \to p_j$, whenever p_i belongs to the strict transform of the exceptional divisor E_j obtained by blowing-up p_j. These valuations were classified by Spivakovsky in [23]. We are interested in the class of exceptional curve valuations (in the terminology of Favre–Jonsson [8]) which corresponds to Case 3 in [23] and to type C from Delgado–Galindo–Núñez [5]. These valuations are characterized by the fact that there exists a point $p_r \in \mathcal{C}$ such that $p_i \to p_r$ for all $i > r$.

Notice that if we consider the surface X_r obtained after blowing-up p_r and the flag

$$E_\bullet := \{X = X_r \supset E_r \supset \{q := p_{r+1}\}\},$$

then the valuation ν is just $\nu_{E_\bullet}$. According to the above mentioned, for an element $f \in R = \mathcal{O}_{\mathbb{P}^2_p}$, we have $\nu(f) = (\nu_1(f), \nu_2(f))$ with $\nu_1(f) = \nu_{E_r}(f)$ and $\nu_2(f) = \mathrm{ord}_q\left(\pi^*(f)/z_r^{\nu_1(f)}\right)$, where $\pi \colon X_r \to X_0$ is the composition of the first r point blowing-ups in η, $z_r = 0$ a local equation for E_r and $\pi^*(f)/z_r^{\nu_1(f)}$ may be seen as a function on E_r. Notice that

$$\nu_2(f) = \left(\pi^*(f)/z_r^{\nu_1(f)}, E_r\right)_q,$$

where $(\cdot\,,\cdot)_q$ denotes the intersection multiplicity at q.

The divisor E_r is defined by a map $\pi \colon X_r \to \mathbb{P}^2$. The intersections of the strict transforms of the exceptional divisors in X_r are represented by the so-called dual graph of π (or of ν_{E_r}). The *geodesic of the dual graph* is defined to be the set of edges (and vertices) in the path joining the vertices corresponding to E_1 and E_r. Additionally, for $i = 1, \ldots, r$, φ_i *will denote an analytically irreducible germ of curve at p whose strict transform is transversal to E_i at a nonsingular point of the exceptional locus.*

In spite of their importance, very few explicit examples of Newton–Okounkov bodies can be found in the literature. We are interested in an explicit computation of the Newton–Okounkov bodies of flags $E_\bullet$ defined by exceptional curve plane valuations ν with respect to the divisor class H given by the pull-back of the line-bundle $\mathcal{O}_{\mathbb{P}^2}(1)$, which we will denote by $\Delta_\nu(H)$. These Newton–Okounkov bodies were studied in [4] for valuations with only one Puiseux exponent [5]. We devote the next section to announce a result which provides an explicit computation of bodies $\Delta_\nu(H)$ for a large class of valuations ν as above which can have an arbitrary number of Puiseux exponents. Its proof and further details will appear elsewhere.

4 The Result

For a start and without loss of generality, we set $(X : Y : Z)$ projective coordinates in $\mathbb{P}^2$, L the line $Z = 0$, which we call the line at infinity, and assume that p is the point with projective coordinates $(1 : 0 : 0)$. Consider also coordinates $x = X/Z$ and $y = Y/Z$ in the affine chart defined by $Z \neq 0$ and local coordinates $u = Y/X$ and $v = Z/X$ around p. With the previous notation, set ν_r a divisorial valuation of the fraction field $K = K(\mathbb{P}^2)$, given by a finite sequence of point blowing-ups $\pi : X_r \to X_0 = \mathbb{P}^2$, whose first blowing-up is at p and is defined by the exceptional divisor E_r. We say that ν_r is *non-positive at infinity* whenever $r \geq 2$, L passes through $p_1 = p$ and p_2 and $\nu_r(f) \leq 0$ for all $f \in \mathbb{C}[x, y] \setminus \{0\}$. Notice that our valuations are valuations centered at infinity [9].

Definition 1 An exceptional curve plane valuation ν of K centered at R is said to be *non-positive at infinity* whenever it is given by a flag

$$E_\bullet := \{X = X_r \supset E_r \supset \{q := p_{r+1}\}\}$$

such that $\nu = \nu_{E_r}$ is non-positive at infinity.

Recall from [7] that the volume of a valuation ν_r as above is defined as

$$\mathrm{vol}(\nu_r) := \lim_{m \to \infty} \frac{\dim_{\mathbb{C}}(R/P_m)}{m^2/2},$$

where $P_m = \{f \in R | \nu_r(f) \geq m\} \cup \{0\}$. Divisorial valuations non-positive at infinity have been studied in [10] and admit an easy characterization:

Theorem 2 *Let ν_r be a divisorial valuation of K centered at p. The valuation ν_r is non-positive at infinity if and only if $\nu_r(v)^2 \geq [\mathrm{vol}\,(\nu_r)]^{-1}$.*

From the previous condition, it is clear that one can find valuations non-positive at infinity with as many Puiseux exponents as one desires. We should also notice the existence of families of surfaces X_r, defined by valuations ν_r as above, whose anticanonical Iitaka dimension is $-\infty$; see [10].

To conclude, we state our result on the Newton–Okounkov bodies $\Delta_\nu(H)$ corresponding to exceptional curve plane valuations non-positive at infinity. Before that, we notice that the proof is based on the fact that Zariski decompositions of certain divisors describe Newton–Okounkov bodies in the case of surfaces [18] and we are able to provide an explicit description of the Zariski decomposition of those divisors, which are $H - tE_r$, where H is the total transform on X_r of a line in $\mathbb{P}^2$ that does not pass through p, and $t \in [0, \nu_r(v)]$.

We will set $\nu_i = \nu_{E_i}$ for $1 \le i \le r$.

Theorem 3 *Let ν be an exceptional curve plane valuation non-positive at infinity and consider its corresponding flag $E_\bullet := \{X = X_r \supset E_r \supset \{q := p_{r+1}\}\}$. Then, the Newton–Okounkov body $\Delta_\nu(H)$ is a triangle; more precisely:*

(I) If $\nu_r(v)^2 > [vol\,(\nu_r)]^{-1}$ then, $\Delta_\nu(H)$ is:

 (i) a triangle with vertices $(0,0)$, $(\nu_r(v), 0)$, $\left(\frac{1}{\mathrm{vol}(\nu_r)\nu_r(v)}, \frac{1}{\nu_r(v)}\right)$, whenever q is a free point in E_r;

 (ii) a triangle with vertices $(0,0)$, $(\nu_r(v), \nu_\ell(v))$, $\left(\frac{1}{\mathrm{vol}(\nu_r)\nu_r(v)}, \frac{\nu_r(\varphi_\ell)}{\nu_r(v)}\right)$, whenever q is a satellite point in $E_\ell \cap E_r$, $\ell < r$ and the vertex given by E_ℓ in the dual graph of ν_r belongs to the geodesic;

 (iii) a triangle with vertices $(0,0)$, $(\nu_r(v), \nu_\ell(v))$, $\left(\frac{1}{\mathrm{vol}(\nu_r)\nu_r(v)}, \frac{\nu_r(\varphi_\ell)+1}{\nu_r(v)}\right)$, otherwise.

(II) If $\nu_r(v)^2 = [\mathrm{vol}\,(\nu_r)]^{-1}$, then $\Delta_\nu(H)$ is:

 (i) a triangle with vertices $(0,0)$, $(\nu_r(v), 0)$, $\left(\nu_r(v), \frac{1}{\nu_r(v)},\right)$, whenever q is a free point in E_r;

 (ii) a triangle with vertices $(0,0)$, $\left(\nu_r(v), \frac{1-\mathrm{vol}(\nu_r)}{\mathrm{vol}(\nu_r)\nu_r(v)}\right)$, $\left(\nu_r(v), \frac{1}{\mathrm{vol}(\nu_r)\nu_r(v)}\right)$, otherwise.

References

1. S.S. Abhyankar, T.T. Moh, Newton–Puiseux expansion and generalized Tschirnhausen transformation I and II. J. Reine Angew. Math. **260**, 47–83 (1973); **261**, 29–54 (1973)
2. S. Boucksom, A. Küronya, C. Maclean, T. Szemberg, Vanishing sequences and Okounkov bodies. Math. Ann. **361**, 811–834 (2015)
3. A. Campillo, O. Piltant, A. Reguera, Curves and divisors on surfaces associated to plane curves with one place at infinity. Proc. Lond. Math. Soc. **84**, 559–580 (2002)
4. C. Ciliberto, M. Farnik, A. Küronya, V. Lozovanu, J. Roé, C. Shramov, Newton–Okounkov bodies sprouting on the valuative tree. Rend. Circ. Mat. Palermo **66**, 161–194 (2017)

5. F. Delgado, C. Galindo, A. Núñez, Saturation for valuations on two-dimensional regular local rings. Math. Z. **234**, 519–550 (2000)
6. M. Dumnicki, B. Harbourne, A. Küronya, J. Roé, T. Szemberg, Very general monomial valuations of $\mathbb{P}^2$ and a Nagata type conjecture. Comm. Anal. Geom. **25**, 125–161 (2017)
7. L. Ein, R. Lazarsfeld, K. Smith, Uniform approximation of Abhyankar valuations in smooth function fields. Am. J. Math. **125**, 409–440 (2003)
8. C. Favre, M. Jonsson, *The Valuative Tree*. Lecture Notes in Mathematics, vol. 1853 (Springer, Berlin, 2004)
9. C. Favre, M. Jonsson, Eigenvaluations. Ann. Sci. Éc. Norm. Sup. **40**, 309–349 (2007)
10. C. Galindo, F. Monserrat, The cone of curves and the Cox ring of a rational surface given by divisorial valuations. Adv. Math. **290**, 1040–1061 (2016)
11. C. Galindo, F. Monserrat, J.J. Moyano-Fernández, Minimal plane valuations. J. Algebraic Geom. **27**, 751–783 (2018)
12. S.-Y. Jow, Okounkov bodies and restricted volumes along very general curves. Adv. Math. **223**, 1356–1371 (2010)
13. K. Kaveh, A. Khovanskii, Newton-Okounkov bodies, semigroups of integral points, graded algebras and intersection theory. Ann. Math. **176**, 925–978 (2012)
14. A. Küronya, V. Lozovanu, C. Maclean, Convex bodies appearing as Okounkov bodies of divisors. Adv. Math. **229**, 2622–2639 (2012)
15. A. Küronya, V. Lozovanu, Local positivity of linear series on surfaces. Algebra Number Theor. **12**, 1–34 (2018)
16. A. Küronya, V. Lozovanu, Positivity of line bundles and Newton–Okounkov bodies. Dec. Math. **22**, 1285–1302 (2017)
17. A. Küronya, V. Lozovanu, Infinitesimal Newton–Okounkov bodies and jet separation. Duke Math. J. **166**, 1349–1376 (2017)
18. R. Lazarsfeld, M. Mustaţă, Convex bodies associated to linear series. Ann. Sci. Éc. Norm. Sup. **42**, 783–835 (2009)
19. A. Okounkov, Brunn-Minkowski inequality for multiplicities. Invent. Math. **125**, 405–411 (1996)
20. A. Okounkov, Note on the Hilbert polynomial of a spherical variety. Funct. Anal. Appl. **31**, 138–140 (1997)
21. A. Okounkov, Why would multiplicities be log-concave?, *The Orbit Method in Geometry and Physics*. Progress in Mathematics, vol. 213 (Berkhauser, Basel, 2003), pp. 329–347
22. J. Roé, Local positivity in terms of Newton Okounkov bodies. Adv. Math. **301**, 486–498 (2016)
23. M. Spivakovsky, Valuations in function fields of surfaces. Am. J. Math. **112**, 107–156 (1990)

Sufficient Conditions for the Finite Generation of Valuation Semigroups

Alex Küronya and Joaquim Roé

Abstract The purpose of this short note is to draw more attention to a very general finite generation problem arising in valutation theory with exciting links to both algebra and geometry. In particular, we propose a few problems with the aim of connecting finite generation in local versus global settings.

A valuation or a set of valuations on a local ring determine a natural semigroup of values, whose algebraic properties are in general difficult to understand. Such semigroups associated to valuations are fundamental objects lying at the crossroads of commutative algebra, combinatorics, and algebraic geometry, their finite generation is an extremely important and equally difficult issue. In our setting the semigroups that arise are most often additive subsemigroups of $\mathbb{N}^n$.

Finite generation properties of closely related 'global' objects such as the semigroup of effective or ample classes in the Néron–Severi group, the section ring of a line bundle, the cone of curves, or the Cox ring, are closely related problems that are at times better understood because of the extra structure and constraints coming from projective geometry. At the same time, surprisingly enough, finite generation verified in the local case can also be used to prove global counterparts; see [9, 10, 16].

A. Küronya
Department of Algebra, Budapest University of Technology and Economics, P.O. Box 91, Budapest 1521, Hungary
e-mail: kuronya@math.uni-frankfurt.de

A. Küronya
Goethe-Universität Frankfurt am Main, Robert-Mayer Str. 6-10, 60325 Frankfurt am Main, Germany

J. Roé (✉)
Departament de Matemàtiques, Universitat Autònoma de Barcelona, 08193 Bellaterra, Barcelona, Spain
e-mail: jroe@mat.uab.cat

© Springer Nature Switzerland AG 2018
M. Alberich-Carramiñana et al. (eds.), *Extended Abstracts February 2016*,
Trends in Mathematics 9, https://doi.org/10.1007/978-3-030-00027-1_2

Recently, the semigroups of values on the graded algebra of a line bundle, or on a Cox ring (direct translations of the local version) and their finite generation have come into focus in [1, 11]; as shown in the latter paper some of the benefits of such results are the existence of completely integrable systems of certain smooth projective varieties.

Sufficient conditions for finite generation have been in existence both in the local case and in the global one. Some of these sufficient conditions coming from the two different settings are apparently analogous to a certain extent (existence of a generating sequence [16] in the local case, existence of maximal divisors [1] in the global case).

In this note we focus on semigroups isomorphic to subsemigroups of $\mathbb{N}^n$ for some n (which for simplicity we call numerical semigroups) with the goal of highlighting the similarities and connections between the known phenomena in local and projective contexts, and propose some problems aiming to a common general approach.

1 Valuation Semigroups in a Local Domain

A valuation on a local domain R is a valuation $\nu\colon K^* \to G$ on its field of fractions which is non-negative on R (i.e., $R \subset R_\nu \subset K$, where R_ν denotes the valuation ring of ν). When R is an algebra over a base field k, valuations are often implicitly assumed to be trivial on k. Given a finite sequence of valuations $\underline{\nu} = (\nu_1, \ldots, \nu_n)$ with value groups $G_1, \ldots, G_n$, the set of tuples of values

$$\Gamma_{\underline{\nu}}(R) = \underline{\nu}(R \setminus \{0\}) \subset G_1 \times \cdots \times G_n \overset{\text{def}}{=} G$$

is a semigroup of great importance (see Teissier's contribution to this volume for an application of finite generation in the case $n = 1$ to local uniformization). If $n = 1$ and the valuation ν_1 is discrete of rank 1, i.e., $G \cong \mathbb{Z}$, then $\Gamma_{\underline{\nu}}(R)$ is a subsemigroup of $\mathbb{N}$ and hence finitely generated, but in virtually all other cases, the basic question whether $\Gamma_{\underline{\nu}}(R)$ is finitely generated becomes really difficult.

A sequence of valuations $\underline{\nu}$ determines a filtration by ideals in R:

$$J_\gamma \overset{\text{def}}{=} \{f \in R \mid \underline{\nu}(f) \geq \gamma\},$$

where $\nu \geq \gamma$ means $\nu_i \geq \gamma_i$ for all i. A *generating sequence* for $\underline{\nu}$ is a finite set $\Lambda = \{\lambda_1, \ldots, \lambda_k\} \subset R$ such that, for every $\gamma \in G$, the ideal J_γ can be generated by a set of products of elements of Λ.

Remark 1 As one would expect, if a generating sequence exists, then $\Gamma_{\underline{\nu}}(R)$ is finitely generated; see [16]. When $n = 1$, there is more to say: consider the second filtration

$$J_{\gamma^+} \overset{\text{def}}{=} \{f \in R \mid \nu(f) > \gamma\}.$$

In this case, a generating sequence immediately gives the finite generation of the multigraded algebra

$$Gr_\nu(R) \overset{\text{def}}{=} \bigoplus_{\gamma \in G} \frac{J_\gamma}{J_{\gamma^+}}.$$

The support of $Gr_\nu(R)$, i.e., the set of degrees γ for which the quotient J_γ / J_{γ^+} is nonzero, is exactly $\Gamma_\nu(R)$. This is always finitely generated if $Gr_\nu(R)$ is. We shall see later that this construction has analogues in semigroups arising from projective geometry. When $n > 1$, the natural graded algebra to consider has graded pieces J_γ / J_{γ^+}, with $J_{\gamma^+} \overset{\text{def}}{=} \{f \in R \mid \underline{\nu}(f) > \gamma\}$ where $\underline{\nu}(f) > \gamma$ means $\nu_i(f) > \gamma_i$ for each i; but in this case the support semigroup may be larger than the value semigroup.

Already the seemingly innocent case of n discrete valuations of rank one on $R = \mathbb{C}[\![x_1, \ldots, x_d]\!]$ shows some of the difficulties that appear, and has received considerable attention in the literature; see [7, 16] and the references therein.

Theorem 2 (Delgado–Galindo–Núñez, [7]) *If $R = k[\![x_1, x_2]\!]$, where k is an algebraically closed field and $\nu_1, \ldots, \nu_n$ are discrete valuations of rank one, then a generating sequence exists.*

In the two-dimensional case, generating sequences are explicitly constructed from the combinatorics of a blowup of the plane where the ν_i become divisorial valuations, or equivalently, as polynomials in x_1, x_2 with prescribed contacts with (usually singular) curve branches customarily attached to the ν_i. In higher dimensions, such generating sequences need not exist; see [16].

2 Valuation Semigroups in Projective Geometry

Let X be a normal projective variety of dimension n over an algebraically closed field k, L a line bundle on X. The finite generation of the section ring

$$R(X; L) \overset{\text{def}}{=} \bigoplus_{m \geq 0} H^0(X, L^{\otimes m})$$

of L is an extremely important question.

For instance, it was a long-standing conjecture, first proved in [3] (see also [5]), that for every smooth projective variety X, the canonical ring $R(X; \omega_X)$ is finitely generated. A discussion of finite generation results for section rings and consequences is provided in [8, §1.8].

One of the most important building blocks of the theory that we would like to mention here is a well-known theorem of Zariski (see [13, 17, §2.1]) which claims that a sufficient condition for the finite generation of L is for L to be semiample, i.e., global sections of some tensor power $L^{\otimes m}$ of L should give rise to a morphism from

X to projective space. More generally, if $L_1, \ldots, L_n$ are semiample line bundles, then the multigraded section ring

$$R(X; L_1, \ldots, L_n) \overset{\text{def}}{=} \bigoplus_{m_i \geq 0} H^0(X, L_1^{\otimes m_1} \otimes \cdots \otimes L_n^{\otimes m_n})$$

is finitely generated.

The *support semigroup*, formed by the multidegrees $(m_1, \ldots, m_n) \in \mathbb{N}^n$ such that

$$H^0(X, L_1^{\otimes m_1} \otimes \cdots \otimes L_n^{\otimes m_n}) \neq 0$$

is finitely generated whenever the graded ring is. A priori, finite generation of the support semigroup is a much weaker (hence presumably easier) condition than that of the ring; however, open questions such as Nagata's celebrated conjecture [13, §5.1.14] boil down to the finite generation of support semigroups of section rings.

We are interested in the semigroups arising when a valuation $\nu\colon K(X)^* \to G$ is additionally given. This determines filtrations

$$F_\gamma H \overset{\text{def}}{=} \{f \in H \mid \nu(f) \geq \gamma\}$$

$$F_{\gamma^+} H \overset{\text{def}}{=} \{f \in H \mid \nu(f) > \gamma\}$$

where H can be $H = H^0(X, L)$, $H = R(X; L)$, $H = R(X; L_1, \ldots, L_n))$, or some similar object. The filtration by the ideals $F_\bullet R(X; L)$ determines a a $\mathbb{N} \times G$-graded k-algebra

$$Gr_\nu R(X; L) \overset{\text{def}}{=} \bigoplus_{\gamma \in G} \frac{F_\gamma R(X; L)}{F_{\gamma^+} R(X; L)} = \bigoplus_{m, \gamma} \frac{F_\gamma H^0(X, L^{\otimes m})}{F_{\gamma^+} H^0(X, L^{\otimes m})}.$$

One situation of specific interest, motivated by the theory of Newton–Okounkov bodies and its applications, is the following.

Question 3 Let X be a projective variety of dimension n, L a big line bundle on X, $\nu\colon \mathbb{C}(X) \to \mathbb{N}^n$ a rank n valuation. When is $Gr_\nu R(X; L)$ finitely generated? Note that in this particular case finite generation of the algebra is equivalent to the finite generation of its support semigroup [14], and the latter can be described as

$$\Gamma_\nu(L) \overset{\text{def}}{=} \bigcup_{m \geq 0} \{m\} \times \nu(H^0(X, L^{\otimes m}) \setminus \{0\}) \subseteq \mathbb{N}^{n+1}.$$

Most literature on Newton–Okounkov bodies focuses on the rank n valuations determined by *admissible flags* on X [14]; we denote $\Gamma_{Y_\bullet}(L)$ the semigroup determined by the valuation associated to the flag $Y_\bullet$.

Note however that, at the cost of changing the birational model we work with, one can reduce the case of general rank n valuations to the admissible flag valuation case by [6, Thm. 2.9].

Remark 4 (*Valuation semigroups and integrable systems*) It was recently pointed out in a work of Harada–Kaveh [11] that finitely generated valuation semigroups lead to completely integrable systems on smooth projective varieties. The basic idea goes as follows: let X be a smooth projective variety, L a very ample line bundle on X. Assume that we can find an admissible flag $Y_\bullet$ on X such that the associated valuation semigroup $\Gamma_{Y_\bullet}(L)$ is finitely generated.

As described in [1], $\Gamma_{Y_\bullet}(L)$ then gives rise to a Gröbner deformation to a (possibly non-normal) toric variety, whose moment map can be then pulled back via the gradient flow to X, which results in a completely integrable system on the original variety X.

Note the two drawbacks of the construction: first, the completely integrable systems arising this way are notoriously difficult to determine (since pulling back by the gradient flow is not an effective algebraic construction), second, the finite generation of $\Gamma_{Y_\bullet}(L)$ is ridiculously difficult to check even in the most concrete cases.

Valuation semigroups tend not to be finitely generated even in the most innocent-looking cases.

Example 5 (*Complete linear series on curves*) Let X be a smooth projective curve of genus g, L an effective line bundle of degree $d > 0$. Every nontrivial valuation on X arises as the order of vanishing at a point of X; let us pick a point $x \in X$. A quick computation using Riemann–Roch [14] (see also [4]) shows that for all $m \in \mathbb{N}$ and all indices $0 \le i \le m \cdot \deg(D) - 2g - 2$ there exists $s \in H^0(X, \mathcal{O}_X(mD))$ such that $\operatorname{ord}_x(s) = i$.

If $D - \deg(D)x$ is *not* a torsion point in $\operatorname{Pic}^0(X)$, then $\deg(D)$ will never be in the image of the normalized valuation map

$$\bigoplus_m H^0(X, \mathcal{O}_X(mD)) \longrightarrow \Delta_{Y_\bullet}(D) = [0, \deg(D)]$$

$$s \longmapsto \frac{\operatorname{ord}_x(s)}{m}.$$

On the other hand, the point $\deg(D)$ lies in the closure of the image, hence $\Gamma_{Y_\bullet}(D)$ cannot be finitely generated.

Remark 6 Note that the above example shows that on a given variety X, finite generation of valuation semigroups is arguably more difficult than the same question for section ring semigroups.

The case of smooth surfaces is already very poorly understood. We propose the following.

Question 7 (Valuation semigroups on surfaces with effective anti-canonical classes)
Let X be a surface with nef or effective anti-canonical class. Find a very ample line
bundle L and an admissible flag $Y_\bullet$ on X such that $\Gamma_{Y_\bullet}(L)$ is finitely generated.

The one known general condition for the finite generation of valuation semigroups
in [2] uses iterated complete intersections, but is unfortunately still not easy to check
in practice.

Theorem 8 *Let X be a nonsingular projective variety of dimension n, and let L be
a very ample divisor. Suppose that, under the embedding $X \hookrightarrow \mathbb{P} = \mathbb{P}(H^0(X, L))$,
there exist linear subspaces $W_n \subseteq W_{n-1} \subseteq \mathbb{P}$, of codimensions n and $n - 1$, respectively, such that the set-theoretic intersection $Y_n = X \cap W_n$ is a single point and the
scheme-theoretic intersection $W_{n-1} \cap X$ is reduced and irreducible.*

*Then, there is a flag $Y_\bullet$, with Y_k an irreducible Cartier divisor in Y_{k-1} for $1 \le k \le
n - 1$, such that the semigroup $\Gamma_{Y_\bullet}(L)$ is finitely generated.*

Nevertheless, a partial answer in the case of some Fano varieties is provided
by [12]. Jow's argument for finite generation goes by verifying Theorem 8 in the
case at hand.

Theorem 9 (S.-Y. Jow, [12]) *Let X be a nonsingular complex projective Fano
variety of dimension n and index $r \ge n - 1$. There exist a very ample divisor L, and
a flag $Y_\bullet$, such that the semigroup $\Gamma_{Y_\bullet}(L)$ is finitely generated.*

Remark 10 Jow shows that by picking $n - 1$ general elements $H_1, \ldots, H_{n-1}$ of the
linear series $|H|$ where $rH = -K_X$ (here again r stands for the index of the Fano
variety X, that is, r is the largest positive integer such that $-K_X$ is divisible by
r in the Néron–Severi group), and a carefully chosen element $H_n \in |H|$ (relying
on the elliptic curve structure of $H_1 \cap \cdots \cap H_{n-1}$), the resulting flag consisting of
the successive complete intersections $H_1 \cap \cdots \cap H_i$ will yield a finitely generated
valuation semigroup.

3   Local-Global Interactions

Each of the finite generation theorems stated above rely for their proofs on exhibiting
(or showing the existence of) elements in a ring that 'follow closely' the defining
data of the valuation(s) one is interested in. In the case of Theorem 2, the data are
given by the centers of the valuations and their proximity combinatorics, and they
allow to construct generating sequences in the local ring R. In the case of Theorems 8
and 9 the data are given by the flag defining the valuation, which is carefully chosen
to be cut by sections of $H^0(X, L)$, so these sections automatically give elements in
$R(X; L)$ that 'follow' the flag, and thus enable inductive arguments for instance.

Question 11 Set up a general framework for rings with $\mathbb{N}^n$-filtrations and sufficient
conditions for finite generation of the resulting semigroups that encompasses the
results known for the local and projective cases.

In addition to such analogies, we want to mention that the generating sequences which work locally to show finite generation of valuation semigroups have also been exploited (implicitly) to show finite generation of section rings on surfaces [10].

On the other hand, the history of valuation semigroups and the related literature is much older and extensive for local rings than for section rings in projective algebraic geometry. One would like to hope for projective analogs of the most recent and sophisticated results for local rings, such as [15], which shows, among other things, that if R is an excellent Noetherian equicharacteristic local domain with algebraically closed residue field, and ν is an Abhyankar zero-dimensional valuation, then there are local domains R' essentially of finite type over R and dominated by the valuation ring R_ν such that the semigroup of values of ν on R' is finitely generated.

A projective analogue of the above would be a weak version of what is needed for the Harada–Kaveh result to take effect.

Question 12 Let X be a normal projective variety over the complex numbers. Can we guarantee the existence of a proper birational model X' of X, a very ample line bundle L and a rank $n = \dim X$ valuation on X' such that $\Gamma_\nu(L)$ is finitely generated?

References

1. D. Anderson, Okounkov bodies and toric degenerations. Math. Ann. **356**(3), 1183–1202 (2013)
2. D. Anderson, A. Küronya, V. Lozovanu, Okounkov bodies of finitely generated divisors. Int. Math. Res. Not. IMRN **2014**(9), 2343–2355 (2014)
3. C. Birkar, P. Cascini, C.D. Hacon, J. McKernan, Existence of minimal models for varieties of log general type. J. Am. Math. Soc. **23**(2), 405–468 (2010)
4. S. Boucksom, A. Küronya, C. Maclean, T. Szemberg, Vanishing sequences and Okounkov bodies. Math. Ann. **361**(3–4), 811–834 (2015)
5. P. Cascini, V. Lazić, New outlook on the minimal model program, I. Duke Math. J. **161**(12), 2415–2467 (2012)
6. C. Ciliberto, M. Farnik, A. Küronya, V. Lozovanu, J. Roé, C. Shramov, Newton–Okounkov bodies sprouting on the valuative tree, arXiv:1602.02074v1 (to appear in the Rendiconti del Circ. Mat. Palermo)
7. F. Delgado, C. Galindo, A. Núñez, Generating sequences and Poincaré series for a finite set of plane divisorial valuations. Adv. Math. **219**, 1632–1655 (2008)
8. T. de Fernex, L. Ein, M. Mustață, Vanishing theorems and singularities in birational geometry (preprint, 2014), pp. 401, http://homepages.math.uic.edu/~ein/DFEM.pdf
9. C. Galindo, F. Monserrat, The cone of curves associated to a plane configuration. Comment. Math. Helv. **80**(1), 75–93 (2005)
10. C. Galindo, F. Monserrat, The cone of curves and the cox ring of rational surfaces given by divisorial valuations. Adv. Math. **290**, 1040–1061 (2016)
11. M. Harada, K. Kaveh, Integrable systems, toric degenerations, and Okounkov bodies, integrable systems, toric degenerations and Okounkov bodies. Invent. Math. **202**(3), 927–985 (2015)
12. S-Y. Jow, Fano varieties with finitely generated semigroups in the Okounkov body construction, arXiv:1511.01197
13. R. Lazarsfeld, *Positivity in Algebraic Geometry. I. Classical Setting: Line Bundles and Linear Series*. Ergebnisse der Mathematik und ihrer Grenzgebiete. 3. Folge. A Series of Modern Surveys in Mathematics, vol. 48 (Springer, Berlin, 2004), p. xviii+387. ISBN: 3-540-22533-1

14. R. Lazarsfeld, M. Mustaţă, Convex bodies associated to linear series. Ann. Sci. Éc. Norm. Supér. (4) **42**(5), 783–835 (2009)
15. B. Teissier, Overweight deformations of affine toric varieties and local uniformization, *Valuation Theory in Interaction*. EMS Series of Congress Report (European Mathematical Society, Zürich, 2014), pp. 474–565
16. L. van Langenhoven, W. Veys, Semigroup and Poincaré series for a finite set of divisorial valuations. Rev. Mat. Complut. **28**, 191–225 (2015)
17. O. Zariski, The theorem of Riemann–Roch for high multiples of an effective divisor on an algebraic surface. Ann. Math. **76**(2), 560–615 (1962)

From Convex Geometry of Certain Valuations to Positivity Aspects in Algebraic Geometry

Victor Lozovanu

Abstract A few years ago Okounkov associated a convex set (Newton–Okounkov body) to a divisor, encoding the asymptotic vanishing behaviour of all sections of all powers of the divisor along a fixed flag. This brought to light the following guiding principle "use convex geometry, through the theory of these bodies, to study the geometrical/algebraic/arithmetic properties of divisors on smooth projective varieties". The main goal of this survey article is to explain some of the philosophical underpinnings of this principle with a view towards studying local positivity and syzygetic properties of algebraic varieties.

One of the earliest known theorems in the history of mathematics is the Pythagorean theorem. Its proof, due to Pythagora (c. 570–c. 495 BC), can be explained by the two pictures in Fig. 1.

Comparing the area of the "white" region in the first picture with the area of the "white" region in the second implies the Pythagorean identity. More importantly, it is captivating how the proof uses convex geometry (Euclidean geometry of polygons) to explore algebraic equations.

This philosophy was revived later in a spectacular but simple fashion by Newton. In a letter to Oldenburg from 1671, Newton has the idea to associate to a polynomial in two variables $f = \sum_{i,j} a_{ij} x^i y^j \in \mathbb{C}[x, y]$ a convex set, called *the Newton polygon*, as follows

$$\mathbb{R}^2 \supseteq \Delta(f) \stackrel{\text{def}}{=} \text{convex hull}\{(i, j) \in \mathbb{Z}^2 \mid a_{ij} \neq 0\},$$

Most of the material presented here is joint work with Alex Küronya. The author would like to thank him for the support, advices and many interesting conversations on the subject. The author would also like to thank the organizers of the "Workshop on Positivity and Valuations" in Barcelona for organizing such a beautiful event and for giving the opportunity to talk about the ideas above.

V. Lozovanu (✉)
Institut für Algebraische Geometrie, Leibniz Universität Hannover, Welfengarten 1, 30167 Hannover, Germany
e-mail: victor.lozovanu@gmail.com

© Springer Nature Switzerland AG 2018
M. Alberich-Carramiñana et al. (eds.), *Extended Abstracts February 2016*,
Trends in Mathematics 9, https://doi.org/10.1007/978-3-030-00027-1_3

Fig. 1 Proof of Pythagora's
theorem

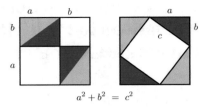

$$a^2 + b^2 = c^2$$

extending the idea of the degree of a polynomial to a polygon. Using the shape of $\Delta(f)$, Newton describes an exhaustive algorithm that finds all the solutions $y = y(x)$ of the equation $f(x, y) = 0$ as Puiseux series. Note that if one thinks of the equation $f(x, y) = 0$ as an affine curve $C \subseteq \mathbb{C}^2$, then this is equivalent to constructing a resolution of singularities of the curve C; see [4, Ch. I] for a nice exposition of these ideas.

After the work of Newton, the above convex construction has slowly found many applications in mathematics. In algebraic geometry it was pioneered by Arnold, who conjectured that it might be possible to express many invariants associated to a holomorphic function in terms of its Newton polygon, at least for "almost all functions" with a given polygon. This has been worked out in the 1970s by Arnold's school (D. Bernstein, A. Khovanskii, A. Kouschnirenko and A. Varchenko).

The idea, due to Khovanskii, is to use the polygon $\Delta := \Delta(f)$ to construct a compactification $X(\Delta)$ of $\mathbb{C}^2$ by adding projective lines at infinity. The projective surface $X(\Delta)$ is special and is called a *toric surface*. Toric varieties in general, originally introduced by Demazure in 1970, are a very special class of projective varieties constructed from convex geometric data (polytopes, cones, fans etc.). Thus many properties/invariants of the toric variety can be translated into the convex geometric world and viceversa. In our case, then studying $f \in \mathbb{C}[x, y]$ is equivalent to doing the same for the divisor $\overline{C} \subseteq X(\Delta)$, the compactification of $C := \{f = 0\}$. Consequently, using the toric language, to study the geometry of $\overline{C}$ is the same as to explore the Euclidean geometry of $\Delta(f)$.

It's worth mentioning here two results that capture nicely not only how can convex geometry be used to study algebraic varieties, but also how to explore convex shapes using algebraic geometry.

Theorem 1 (i) *(Khovanskii) Let $C \subseteq \mathbb{C}^2$ be an irreducible curve defined by a polynomial $f \in \mathbb{C}[x, y]$. Then the topological genus $g(C)$ is at most the number of integral points contained in the interior of the Newton polygon $\Delta(f)$. Equality happens when f is chosen to be generic with a fixed Newton polygon.*

(ii) *(Bernstein) If $\Delta \subseteq \mathbb{R}^2$ is a planar polygon, then the number of integral points contained in $p \cdot \Delta$ ($p \in \mathbb{N}$) is a polynomial function in p (equal to the Euler characteristic $\chi(X(\Delta), p\overline{C})$).*

1 Newton–Okounkov Bodies

Based on the above exposition, it becomes important to know if there is a bridge allowing to explore any projective variety using convex geometry. Note that the Newton polygon is constructed from a very specific valuation, which associates to a monomial $x^i y^j$ the vector $(i, j) \in \mathbb{R}^2$. Since the picture is local, then it can be said that the Newton polygon $\Delta(f)$ encodes how all the monomials appearing in f vanish along a fixed set of local parameters.

In order to introduce such a convex construction in the global setting one needs to make some changes to the initial construction. First, instead of a local system of parameters one will consider a complete flag on the initial variety. Instead of associating convex sets to functions we will do that for divisors (or line bundles). And lastly and more importantly, due to the fact that we work in the global setting, we will need to take the asymptotic version of Newton polygon.

Based on these ideas and inspired by work of Khovanskii from the 1970s, Okounkov [12, 13] explained in passing how to associate to an ample divisor a convex set, called *the Newton–Okounkov body*, encoding how all the sections of all powers of the divisor vanish along a fixed flag. The foundation of this theory was then laid down in whole generality by Lazarsfeld–Mustață [10] and independently by Kaveh–Khovanskii [3].

We introduce the construction of Newton–Okounkov bodies in dimension two. The higher-dimensional counterpart is done accordingly in an inductive manner.

Let X be a smooth projective surface, and H a Cartier divisor on X. Let (C, x) be a flag on X, consisting of an irreducible curve $C \subseteq X$ and $x \in C$ a smooth point. We will denote by $\sim$ the linear equivalence of divisors. Then to measure how an effective divisor $D \sim H$ vanishes along the flag (C, x), one constructs a valuation vector $\nu_{(C,x)}(D) = (\nu_1(D), \nu_2(D))$ given by $\nu_1(D) = \text{ord}_C(D)$ and $\nu_2(D) = \text{ord}_x((D - \nu_1(D)C)|_C)$. Finally, the *Newton–Okounkov body* of this data is defined via

$$\Delta_{(C,x)}(H) \overset{\text{def}}{=} \text{closed convex hull}\left(\bigcup_{m \geq 1} \frac{1}{m}\{\nu_{(C,x)}(D) | D \sim mH\} \right) \subseteq \mathbb{R}^2 .$$

If $\vec{v}_x \in T_x X$ is a tangency direction at x, then one can associate the *infinitesimal Newton–Okounkov body* $\Delta_{(x,\vec{v}_x)}(H)$ in a similar way. In particular, it can be seen as the Newton–Okounkov body of H defined by a flag on the blow-up of X at point x.

Since we work in the global setting and take the asymptotic view, it turns out that these convex sets satisfy very good properties.

Properties 2 *(i) The sets $\Delta_{(C,x)}(H)$, $\Delta_{(x,\vec{v}_x)}(H) \subseteq \mathbb{R}^2$ are always compact and closed. On surfaces they turn out to be always polygons, but they can be badly shaped in higher-dimensions.*

(ii) $\Delta_{(C,x)}(pH) = p \cdot \Delta_{(C,x)}(H)$ for any integer $p > 0$. In particular, one can define the Newton–Okounkov body for a $\mathbb{Q}$-divisor and by continuity for any $\mathbb{R}$-divisor.

(iii) We have $H \equiv_{\mathbb{R}} H'$ if and only if $\Delta_{(C,x)}(H) = \Delta_{(C,x)}(H') \ \forall (C, x)$ on X. The first condition also implies that $\Delta_{(x,\vec{v}_x)}(H) = \Delta_{(x,\vec{v}_x)}(H') \ \forall (x, \vec{v}_x) \in X \times T_x X$, but not viceversa.

(iv) $\displaystyle\int_{\Delta_{(C,x)}(H)} 1 \, dt dy = \frac{1}{2} vol_X(H) \Big(:= \lim_{m \to \infty} \frac{h^0(X, mH)}{m^2} \Big)$, for any (C, x).

(v) If $f \in \mathbb{C}[x, y]$ is a non-zero polynomial with Newton polygon Δ, then there exists a flag $(T\text{-invariant})$ on $X(\Delta)$ with respect to which the Newton–Okounkov polygon of $\overline{C}$ is a translate of Δ.

2 Convexity Properties of Asymptotic Invariants

One of the first main applications of Newton–Okounkov bodies, pioneered by Okounkov in [12, 13], is that this theory explains well why some asymptotic invariants in algebraic geometry satisfy properties that seem to come from a convex geometric world. This is well explained by the following example.

Example 3 (Hodge index theorem) Let X be a smooth projective surface, H an ample class on X and $C \subseteq X$ an irreducible curve with $C^2 > 0$. Let $x \in C$ be a smooth point. Then, by Serre vanishing and the definition of Newton–Okounkov polygons, we have

$$O = (0, 0), \quad A = (0, (H.C)) \in \Delta_{(C,x)}(H) .$$

By the same token, the line l of slope $\alpha = -(C^2)$ and passing through A is a supporting line for the polygon $\Delta_{(C,x)}(H)$. Thus, by convexity one has the following inclusion

$$\Delta_{(C,x)}(H) \subseteq \triangle OAB , \quad \text{where } B = \Big(\frac{(H.C)}{C^2}, 0 \Big) .$$

Computing the areas of both polygons we get the inequality $(C^2) \cdot (H^2) \leq (H.C)^2$, i.e., the Hodge index theorem.

The basic principle, underlined in the example, is to use Newton–Okounkov bodies to deduce Hodge index theorem and related statements as a consequence of Brunn–Minkowski inequality for volumes of convex sets. In his ground-breaking proof of log-concavity property of the coefficients of the chromatic polynomial of a graph, Huh [2] gives in passing a simple proof of a higher-dimensional version of Hodge index theorem. Using the same idea, one finds in [3, 10] a short proof of the log-concavity property of the volume function of divisors. Furthermore, in [10] the authors also found an elegant proof of Fujita approximation of the volume of a divisor based solely on semigroup theory and convex geometry.

It is worth mentioning that this theory has rendeered clear how to prove such results in the arithmetic world, which are more difficult to tackle than in complex geometry. For example, Chen [1] uses a probabilistic and convex geometric approach to give a unified proof of a stronger Hodge index inequality both in the algebraic geometry world (over the complex numbers) and the arithmetic geometry one (over algebraic number fields).

3 Local Positivity of Divisors Through Convex Geometry

In order to obtain other applications of Newton–Okounkov bodies in algebraic geometry one doesn't need to dig that deep. Going back to property 3 above, we know that the collection of all of them for a fixed divisor serves as an universal numerical invariant. Since many interesting properties of the divisor, like positivity for example, are numerical in essence, it becomes natural to ask if they can be translated in the language of Newton–Okounkov bodies.

This philosophy is fully carried out by A. Küronya and the author in [5–7]. The goal was to understand how to translate positivity properties of the divisor on any algebraic variety, such as ampleness and nefness, into the convex geometry realm.

But before explaining this, we need to introduce some notation. For positive real number $\lambda, \xi > 0$ set

$$\Delta_\lambda \overset{\text{def}}{=} \{(t, y) \in \mathbb{R}^2_+ \mid t + y \le \lambda\} \quad \text{and} \quad \Delta^*_\xi \overset{\text{def}}{=} \{(t, y) \in \mathbb{R}^2_+ \mid 0 \le t \le \xi, 0 \le y \le t\};$$

see Fig. 2. With this at hand, the two dimensional case is explained by the following theorem:

Theorem 4 *Let X be a smooth projective surface and H an $\mathbb{R}$-divisor on X. Then,*

(i) (Local) H is nef $\Leftrightarrow$ $\forall x \in X$ there exists a flag (C, x) such that $(0, 0) \in \Delta_{(C,x)}(H)$.

   H is ample $\Leftrightarrow$ $\forall x \in X \ \exists (C, x)$ and $\lambda > 0$ such that $\Delta_\lambda \subseteq \Delta_{(C,x)}(H)$.

(ii) (Inf.) H is nef $\Leftrightarrow$ $\forall x \in X \ \exists \vec{v}_x \in T_x X$ such that $(0, 0) \in \Delta_{(x,\vec{v}_x)}(H)$.

   H is ample $\Leftrightarrow$ $\forall x \in X \ \exists \vec{v}_x \in T_x X$ and $\xi > 0$ suc that $\Delta^*_\xi \subseteq \Delta_{(x,\vec{v}_x)}(H)$.

Remark 5 The local picture generalizes to any algebraic smooth surface a combinatorial characterization of T-invariant ample/nef divisors on toric surfaces given in terms of classical Newton polygon. Moreover, the condition of ampleness in the local setting can be seen as an intermediate criteria sitting between the classical cohomological version of Serre and the numerical criterion of Nakai–Moishezon–Kleiman (see [9, Ch. 1] for a nice introduction to this classical material). On the other hand, the fact that positivity of divisors can be detected from information given by tangency direction is completely new in the literature.

Fig. 2 The local and infinitesimal picture of ampleness

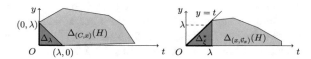

Remark 6 The infinitesimal picture holds also in higher dimensions, as proved by Küronya and the author in [7]. On the other hand, the local picture seems to be harder to deal with and we attempt in [6] to give a weaker version of Theorem 4(i).

Since ampleness can be seen through the convex geometry of Newton–Okounkov bodies, then we are naturally lead to asking whether it is possible to read off numerical invariants of the divisor from these convex sets. A classical invariant, measuring how many jets are asymptotically separated by the divisor at a fixed point, is the *Seshadri constant*. It was introduced by Demailly in the 90s in his work on Fujita conjecture, and is defined to be

$$\epsilon(H;x) \stackrel{\text{def}}{=} \sup_{x \in C \subseteq X} \frac{(H.C)}{\text{mult}_x(C)},$$

where $x \in X$ is a point and H is an ample divisor on X. One of the main results of [5, 7] in higher dimensions) is that the Seshadri constant can be seen on any infinitesimal Newton–Okounkov body defined at the base point x.

Theorem 7 *If H is an ample $\mathbb{R}$-divisor on X and $(x, \vec{v}_x) \in X \times T_x X$ then,*

$$\epsilon(H;x) = max\{\xi > 0 \mid \Delta_\xi^* \subseteq \Delta_{(x,\vec{v}_x)}(H)\}.$$

In particular, the right-hand side does not depend on the tangency direction.

Remark 8 It is worth noting here that the volume of the divisor H and its largest asymptotic multiplicity can also be seen on a fixed infinitesimal Newton–Okounkov body. This in turn leads to the following natural question: Is it possible to translate questions about one invariant to easier to handle questions of another invariant that can be read off from the same convex set?

In a project with A. Küronya and C. Maclean we tackle this issue. If H is an ample Cartier divisor on a surface X and $x \in X$ is a point, then there exists a three-fold Y and another Cartier divisor B on Y such that

$$\epsilon(H;x) \in \mathbb{Q} \iff vol_Y(B) \in \mathbb{Q}.$$

Rationality of Seshadri constants is an old, folklore, and still open question. The equivalence reduces this to rationality of the volume of a Cartier divisor. On the other hand, if either the associated ring of the divisor B is finitely generated or there exists a Zariski decomposition, then the volume has to be rational. Thus our initial question can be linked to problems arising in birational geometry, where much has been developed on the problem of finitely generatedness of the canonical ring.

It has long been known that a divisor has better local positivity properties at a very generic choice of the base point oppose to what is happening at an arbitrary one. For example, an interesting result of Ein–Lazarsfeld from the 90s says that for an ample Cartier divisor H, the Seshadri constant $\epsilon(H;x) \geq 1$ when $x \in X$ is very generic (outside of countably many curves on X). If one considers an arbitrary choice of the point, then the Seshadri constant can take on arbitrarily small positive rational values.

So, it becomes natural to ask whether anything more can be said about the shape of infinitesimal Newton–Okounkov bodies of an ample Cartier divisor at very generic points. This has been tackled in [5] and can be philosophically explained as follows:

Theorem 9 *Let X be a projective surface and H an ample Cartier divisor on X. Let $x \in X$ be a very generic point and $\vec{v}_x \in T_x X$ a generic tangency direction. Then, either we have the inclusion $\Delta_{(x,\vec{v}_x)}(H) \subseteq \triangle OAB$, where $O = (0,0)$, $A = (\epsilon, \epsilon)$ and $B = (\mu, 0)$ with $|\mu - \epsilon| \ll 1$, or there exists a curve $C \subseteq X$ smooth at x with degree (H, C) being small.*

Remark 10 Using the Euclidean volumes of the two convex shapes in Theorem 9 one deduces very strong conditions for lower bounds on Seshadri constants at very generic point. For example, Szemberg has conjectured that for a generic surface $X \subseteq \mathbb{P}^3$ of degree $N \geq 5$ these lower bounds depend on the primitive solution to the Pell's equation $p^2 - q^2 N = 1$. So, using Theorem 9 one can prove Szemberg conjecture for infinitely many choices of N. This and other related problems are tackled in a joint work of the author with A. Küronya and F. Bastianelli.

4 Divisors with Nice Singularities and Convex Geometry

In many groundbreaking results in the last thirty years or so, ranging from diophantine approximation to birational geometry, from Kähler to projective geometry, one important step is the ability to find effective divisors that have nice singularities at a given point. In algebraic geometry these ideas have been pioneered by Mori, Kollar, Demailly, Siu and others and is one of the most powerful techniques in modern algebraic geometry.

On the other hand, going back to the definition of Newton–Okounkov bodies, we know that they encode how all the sections for any power of the divisor vanish along a fixed flag. Thus, one can hope that one might be able to obtain criteria for finding divisors with nice singularities just by looking at these convex sets. Let $\Lambda \stackrel{\text{def}}{=} \{(t, y) \in \mathbb{R}^2 | t \geq 2, t \geq 2y \geq 0\}$. In [8] we give a strong criteria for finding divisors with "nice" singularities in terms of convex geometry of the infinitesimal Newton–Okounkov bodies.

Theorem 11 *Let H be an ample $\mathbb{Q}$-divisor on a smooth projective surface X. If*

$$\text{interior of } \left(\Delta_{(x,\vec{v}_x)}(H) \cap \Lambda \right) \neq \varnothing, \quad \forall \vec{v}_x \in T_x X$$

(see Fig. 3) then, there exists an effective $\mathbb{Q}$-divisor $D \equiv H$ with "nice" singularities at x, i.e., the multiplier ideal $\mathscr{J}(X; D)$ is equal locally (around x) to the maximal ideal of the point x.

Fig. 3 Existence of divisors
with "nice" singularities

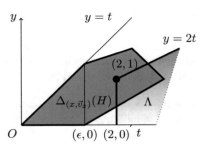

Taking into account the statement of Theorem 11, suppose that we are not able to find a divisor with "nice" singularities. Then there exists a tangency direction $\vec{v}_x \in T_x X$ for which we know that the Newton–Okounkov polygon $\Delta_{(x,\vec{v}_x)}(H)$ sits above a line passing through the point (2, 1). Due to convexity and Theorem 7, this implies that the Seshadri constant $\epsilon(H; x)$ is quite small.

On the other hand, if H is Cartier and $x \in X$ a very generic point, then we can apply Theorem 9. So, either $\epsilon(H; x)$ is large or there exists a curve $C \subseteq X$ smooth at x and $(H.C)$ small. This idea was used in [8] to deduce a nice geometric criteria for finding effective divisors with "nice" singularities at very generic points.

Corollary 12 *Let X be a smooth surface and H an ample Cartier divisor on X with $(L^2) \geq 5(p + 2)^2 (p \in \mathbb{N})$. If $x \in X$ is a very general point and there is no irreducible curve $C \subseteq X$ smooth at x with $1 \leq (L \cdot C) \leq p + 2$, then there exists an effective $\mathbb{Q}$-divisor $D \equiv \frac{1}{p+2} L$ with "nice" singularities as in Theorem 11.*

This kind of statements are important, since it translates the condition of existence of divisors with "nice" singularities to a geometric one, the non-existence of curves of small degree. Furthermore, its applications should be manifold. Below we give a surprising application to the study of syzygies on abelian surfaces.

5   Convex Geometry and Syzygies on Abelian Surfaces

As in the proof of Pythagorean theorem, where one uses convex geometry to explore algebraic equations, we close the circle by explaining how all the ideas above can be used to understand the syzygies on abelian surfaces.

In order to understand algebraically a subvariety $X \subseteq \mathbb{P}^n$ one studies the syzygies of the ideal sheaf $\mathscr{I}_{X|\mathbb{P}^n}$. The first syzygy encodes the generators of this ideal, the second one the relationships between these generators etc. Green has introduced a way of how to see the simplest form of syzygies through properties N_p. For example, N_0 says that the embedding $X \subseteq \mathbb{P}^n$ is projectively normal, N_1 - the ideal sheaf $\mathscr{I}_{X|\mathbb{P}^n}$ is generated by quadrics, N_2 - the relationships between these quadrics are only of linear form etc. (see [9, Ch. 1.8] for a nice introduction on this subject).

Based on all the ideas explained in this survey, in [8] it was found a very interesting criteria for property N_p to be satisfied for abelian surfaces in terms of non-existence of elliptic curves of small degrees. In particular, we build a bridge between the algebraic and geometric world of an abelian surface, and convex geometry plays a paramount role.

Theorem 13 *Let $p \geq 0$ be a natural number, X a complex abelian surface, and L an ample line bundle on X with $(L^2) \geq 5(p+2)^2$. Then the following are equivalent:*

(a) X does not contain an elliptic curve C with $(C^2) = 0$ and $1 \leq (L \cdot C) \leq p+2$;
(b) the line bundle L satisfies property N_p.

Remark 14 It is proved in [11] that on an abelian space property N_p is implied by the existence of a divisor $D = \frac{1}{p+2}L$ with just a bit "nicer" singularitiess as those seen in Theorem 11, i.e., the multiplier ideal $\mathscr{J}(X, D)$ is globally (and not only locally) the maximal ideal of the origin $0 \in X$. Philosophically, Theorem 13 follows by refining the tools used in the proof of Corollary 12.

References

1. H. Chen, Inegalité d'indice de Hodge en géométrie et arithmétique: une approche probabiliste (2014)
2. J. Huh, Milnor numbers of projective hypersurfaces and the chromatic polynomial of graphs. J. Am. Math. Soc. **25**(3), 907–927 (2012)
3. K. Kaveh, A.G. Khovanskii, Newton–Okounkov bodies, semigroups of integral points, graded algebras and intersection theory. Ann. Math. (2) **176**(2), 925–978 (2012)
4. J. Kollár, *Lectures on Resolution of Singularities*, vol. 166. Annals of Mathematics Studies (Princeton University Press, Princeton, 2007), p. vi+208
5. A. Küronya, V. Lozovanu, Local positivity of linear series on surfaces (2014), arXiv:1411.6205
6. A. Küronya, V. Lozovanu, Positivity of line bundles and Newton–Okounkov bodies (2015), arXiv:1506.06525
7. A. Küronya, V. Lozovanu, Infinitesimal Newton–Okounkov bodies and jet separation (2015), arXiv:1507.04339
8. A. Küronya, V. Lozovanu, A Reider-type theorem for higher syzygies on Abelian surfaces (2015), arXiv:1509.08621
9. R. Lazarsfeld, *Positivity in Algebraic Geometry I–II*. Ergebnisse der Mathematik und ihrer Grenzgebiete. 3. Folge, vol. 48–49 (Springer, Berlin, 2004)
10. R. Lazarsfeld, M. Mustaţă, Convex bodies associated to linear series. Ann. Sci. Éc. Norm. Supér. (4) **42**(5), 783–835 (2009)
11. R. Lazarsfeld, G. Pareschi, M. Popa, Local positivity, multiplier ideals, and syzygies of Abelian varieties. Algebra Number Theory **5**(2), 185–196 (2011)
12. A. Okounkov, Brunn–Minkowski inequality for multiplicities. Invent. Math. **125**(3), 405–411 (1996)
13. A. Okounkov, Why would multiplicities be log-concave?, Progress in Mathematics, vol. 213 (Birkhäuser, Boston, 2003), pp. 329–347

Non-positive at Infinity Valuations

Francisco Monserrat

Abstract This note is based on published results in Galindo, Monserrat (Adv Math 290:1040–1061, 2016, [10]) and it is an extended summary of the talk, given by the author, at the Workshop on Positivity and Valuations, held from 22 to 26, February, at Centre de Recerca Matemàtica (Barcelona). We consider surfaces X defined by plane divisorial valuations ν of the quotient field of the local ring R at a closed point p of the projective plane $\mathbb{P}^2$ over an arbitrary algebraically closed field k and centered at R. We characterize those valuations ν which are non-positive (resp., negative) on $\mathcal{O}_{\mathbb{P}^2}(\mathbb{P}^2 \setminus L) \setminus k$, where L is a certain line containing p. Also, under these conditions, we characterize when the Cox ring of X is finitely generated (as k-algebra).

1 Introduction

Along this note, k will denote an algebraically closed field of arbitrary characteristic, $\mathbb{P}^2 := \mathbb{P}^2_k$ the projective plane over k and our valuations will be of the quotient field of the local ring $R := \mathcal{O}_{\mathbb{P}^2, p}$, where p is a fixed point in $\mathbb{P}^2$. To fix notation, we set $(X : Y : Z)$ projective coordinates in $\mathbb{P}^2$, consider the line L with equation $Z = 0$, that will be called the line at infinity, and the point p with projective coordinates $(1 : 0 : 0)$. In addition, pick affine coordinates $x = X/Z$; $y = Y/Z$ in the chart of $\mathbb{P}^2$ given by $Z \neq 0$ and consider a divisorial valuation ν of the quotient field of the local ring R and centered at R. Set $\mathfrak{m}$ the maximal ideal of R; assume that ν is not the $\mathfrak{m}$-adic valuation (that given by $\nu(f) = s$ if and only if $f \in \mathfrak{m}^s \setminus \mathfrak{m}^{s+1}$) and L is the tangent line of ν; see Sect. 2. These conditions are assumed for all valuations we consider. The main goal of this paper is either to characterize the fact

---

F. Monserrat (✉)
Institut Universitari de Matemàtica Pura i Aplicada, Universitat Politècnica de València,
Camino de Vera s/n, 46022 València, Spain
e-mail: framonde@mat.upv.es

© Springer Nature Switzerland AG 2018
M. Alberich-Carramiñana et al. (eds.), *Extended Abstracts February 2016*,
Trends in Mathematics 9, https://doi.org/10.1007/978-3-030-00027-1_4

that the valuation ν is non-positive or negative on all polynomials $f(x, y)$ in the set $k[x, y] \setminus k$. Some of these characterizations involve global geometric properties of the surface X determined by ν.

In [6], for toric varieties, Cox introduced the Cox rings. He showed that these varieties behave like a projective space in many ways. His definition was extended to varieties with free, finitely generated Picard group [11] and, roughly speaking, this ring is the graded one of the section of line bundles on the variety. Finite generation of Cox rings achieves great importance in the minimal model program, since for varieties with this property the mentioned program can be carried out for any divisor. Note that, recently, the existence of minimal models for complex varieties of log general type has been proved [2] and that the Cox ring of a Fano complex variety is finitely generated. With respect to some rational surfaces, the fact that the Cox ring is finitely generated is related to invariant theory and the Hilbert's fourteen problem as one can see in [5, 13, 16]. The recent literature contains a number of papers concerning this issue [1, 8, 9, 12, 14, 15] and confirms that the classification of rational surfaces (and, of course, of varieties) with finitely generated Cox ring is a difficult problem.

Our forthcoming Theorem 2 (resp., Theorem 3) characterizes those valuations that are non-positive (resp., negative) on $k[x, y] \setminus k$. These valuations define surfaces such that the finite generation of their Cox rings can be determined by the conditions we give in Theorem 4.

Finally, we provide examples of families of surfaces, with arbitrarily large Picard number, whose Cox ring is finitely generated and whose anticanonical Iitaka dimension is equal to $-\infty$.

2 Preliminaries

Keep the notations and assumptions given in the Introduction, set ν a divisorial valuation of the quotient field of $R = \mathcal{O}_{\mathbb{P}^2, p}$ centered at R and

$$\pi: X = X_m \xrightarrow{\pi_m} X_{m-1} \longrightarrow \cdots \longrightarrow X_1 \xrightarrow{\pi_1} X_0 = \mathbb{P}^2_k \qquad (1)$$

the simple sequence of point blowing-ups that ν defines. Here π_1 is the blowing-up of $\mathbb{P}^2_k$ at p_1 and π_{i+1}, $1 \leq i \leq m - 1$, the blowing-up of X_i at the unique point p_{i+1} of the exceptional divisor defined by π_i, E_i, such that ν is centered at the local ring $\mathcal{O}_{X_i, p_{i+1}}$. Denote by $C_\nu := \{p_i\}_{i=1}^m$ the sequence (or configuration) of infinitely near points above defined.

The (unique) line that passes through $p = p_1$ and its strict transform goes through p_2 will be called *tangent line* of ν. Throughout this note we will assume that $m \geq 2$ and the tangent line of ν coincides with the line at infinity L (defined by the equation $Z = 0$). Notice that the condition $m \geq 2$ is equivalent to say that ν is not the $\mathfrak{m}$-adic valuation, where $\mathfrak{m}$ denotes the maximal ideal of R.

Definition 1 A valuation ν as above is *non-positive* (resp., *negative*) *at infinity* if $\nu(f) \leq 0$ (resp., $\nu(f) < 0$) for all $f \in k[x, y] \setminus k$.

One of the objectives of this note is to give several characterizations of the non-positive and negative valuations at infinity. To this purpose, we need several definitions and notations.

The volume of ν is defined as

$$\text{vol}(\nu) = \limsup_{\alpha \to \infty} \frac{\text{length}(R/\mathcal{P}_\alpha)}{\alpha^2/2},$$

where $\mathcal{P}_\alpha = \{f \in R | \nu(f) \geq \alpha\} \cup \{0\}$; see [7].

Set $\text{Pic}(X)$ the Picard group of the surface X and $\text{Pic}_{\mathbb{Q}}(X) = \text{Pic}(X) \otimes_{\mathbb{Z}} \mathbb{Q}$ the corresponding vector space over the field of rational numbers. It is well-known that the intersection form extends to a bilinear pairing over $\text{Pic}_{\mathbb{Q}}(X)$. Denote by E_0 a line in $\mathbb{P}^2$ that does not pass through the point p and set $\{E_i\}_{i=0}^m$ (resp., $\{E_i^*\}_{i=0}^m$) the strict (resp., total) transforms of the line E_0 and the exceptional divisors (also denoted $\{E_i\}_{i=1}^m$) on the surface X through π. Denote $[E_i]$ (respectively, $[E_i^*]$), $0 \leq i \leq m$, the class modulo linear (or numerical) equivalence on $\text{Pic}_{\mathbb{Q}}(X)$ of the mentioned divisors. Then $\{[E_i]\}_{i=0}^m$ and $\{[E_i^*]\}_{i=0}^m$ are bases of the vector space $\text{Pic}_{\mathbb{Q}}(X)$ and $E_i = E_i^* - \sum_{p_j \to p_i} E_j^*$ gives a change of basis in the $\mathbb{Q}$-vector space of divisors with exceptional support. A different basis of $\text{Pic}_{\mathbb{Q}}(X)$ is $\{[\tilde{L}]\} \cup \{[E_i]\}_{i=1}^m$, where $\tilde{L}$ denotes the strict transform of L on X and $[\tilde{L}]$ its class in $\text{Pic}_{\mathbb{Q}}(X)$.

Consider the dual basis of $\{[\tilde{L}]\} \cup \{[E_i]\}_{i=1}^m$, that is, the one given by classes of divisors $D_0, D_1, \ldots, D_m$ on X such that $D_0 \cdot L = 1$, $D_0 \cdot E_i = 0$ and $D_i \cdot E_j = \delta_{ij}$ for all $i, j \in \{1, \ldots, m\}$, where δ_{ij} denotes the Kronecker delta. Notice that we can take $D_0 = E_0^*$ and

$$D_i := d_i E_0^* - \sum_{j=1}^m \text{mult}_{p_j}(\varphi_i) E_j^*,$$

$1 \leq i \leq m$. In this definition, on the one hand, d_i denotes the intersection multiplicity at p $d_i = (\varphi_L, \varphi_i)_p$, φ_L being the germ of the line L at p, and φ_i being an analytically irreducible germ of curve at p whose strict transform on X_i is transversal to E_i at a non-singular point of the exceptional locus. On the other hand, $\text{mult}_{p_j}(\varphi_i)$ denotes the multiplicity at p_j of the strict transform of φ_i at p_j.

The *cone of curves* of X, $NE(X)$, is defined as the convex cone of $\text{Pic}_{\mathbb{Q}}(X)$ generated by the classes in $\text{Pic}_{\mathbb{Q}}(X)$ of effective divisors on X.

Another objective of this note is to show conditions that characterize the finite generation of the Cox ring of the surfaces X associated with non-positive valuations at infinity. Next, we recall the definition of Cox ring.

Denote $\mathbf{s} = (s_0, s_1, \ldots, s_m) \in \mathbb{Z}^{m+1}$, $T_{\mathbf{s}} := \sum_{i=0}^m s_i E_i$ and regard the vector spaces

$$H^0(X, \mathcal{O}_X(T_{\mathbf{s}})) = \{f \in k(X) \setminus \{0\} \mid \text{div}_X(f) + T_{\mathbf{s}} \geq 0\} \cup \{0\}$$

as k-vector subspaces of the function field $k(X)$ of X. The *Cox ring* of X is defined as

$$\mathrm{Cox}(X) := \bigoplus_{\mathbf{s} \in \mathbb{Z}^{m+1}} H^0\left(X, \mathcal{O}_X(T_{\mathbf{s}})\right),$$

where we must notice that different bases of $\mathrm{Pic}(X)$ give isomorphic (as k-algebras) Cox rings.

3   The Results

Our first result provides three characterizations of the non-positive at infinity valuations, two of them involving the volume and the cone of curves.

Theorem 2 ([10, Th. 1]) *Let ν be a plane divisorial valuation of the quotient field of R centered at R. Set X the surface that it defines via its attached sequence of point blowing-ups π (1). Assume that the number m of blowing-ups is at least 2 and the tangent line of ν is the line at infinity L. Then, the following conditions are equivalent:*

(a) *ν is non-positive at infinity;*
(b) *$d_m^2 \geq \mathrm{vol}(\nu)^{-1}$;*
(c) *D_m is a nef divisor;*
(d) *the cone $NE(X)$ is spanned by $\{[\tilde{L}]\} \cup \{[E_i]\}_{i=1}^m$.*

Next, we characterize the negative at infinity valuations. But first, we need to recall the notion of Iitaka dimension, $\kappa(D)$, of a divisor D on X. It is defined to be

$$\kappa(D) := \max\{\dim \varphi_{|nD|}(X)\},$$

where n runs over the set $\{m \in \mathbb{Z}_{>0} \mid H^0(X, \mathcal{O}_X(mD)) \neq 0\}$, $\mathbb{Z}_{>0}$ denotes the set of positive integers, dim projective dimension and, for each n, $\varphi_{|nD|}(X)$ is the closure of the image of the rational map $\varphi_{|nD|} \colon X \cdots \to \mathbb{P}H^0(X, \mathcal{O}_X(nD))$ defined by the complete linear system $|nD|$. By convention, $\kappa(D) = -\infty$ whenever $|nD| = \emptyset$ for all $n > 0$ and it holds that either $\kappa(D) = -\infty$ or $0 \leq \kappa(D) \leq \dim(X)$.

Theorem 3 ([10, Th. 2]) *Keep the same assumptions and notations as in Theorem 2. Then, the following conditions are equivalent:*

(a) *ν is negative at infinity;*
(b) *either $d_m^2 > \mathrm{vol}(\nu)^{-1}$, or $d_m^2 = \mathrm{vol}(\nu)^{-1}$ and $\kappa(D_m) = 0$;*
(c) *$D_m \cdot \tilde{C} > 0$ for any integral curve C on $\mathbb{P}^2$ different from L, $\tilde{C}$ being the strict transform on X of the curve C.*

Moreover, when the characteristic of k is zero, condition (b) *can be replaced by*

(b') *either $d_m^2 > \mathrm{vol}(\nu)^{-1}$, or $d_m^2 = \mathrm{vol}(\nu)^{-1}$ and $\dim |D_m| = 0$.*

Finally, we will determine which ones of the non-positive at infinity valuations provide surfaces X with finitely generated Cox ring.

Theorem 4 ([10, Cor. 4]) *Let X be a surface defined by a plane divisorial valuation ν as in Theorem 2. Assume that the equivalent statements given in that theorem happen. Then, $\mathrm{Cox}(X)$ is a finitely generated k-algebra if, and only if, for all $i \in \{2, \ldots, m\}$, either $D_i^2 > 0$, or $D_i^2 = 0$ and $\kappa(D_i) > 0$. If the characteristic of the field k is zero, the condition $\kappa(D_i) > 0$ can be replaced by $\dim |D_i| > 0$.*

The above result provides a wide range of rational surfaces with finitely generated Cox ring. Moreover their anticanonical Iitaka dimension can be $-\infty$, as the next example shows. It gives a new infinite family of surfaces with both conditions and with arbitrarily large Picard number.

Example 5 Assume that the characteristic of the field k is zero. Fix two positive integers a and r such that $r \geq a \geq 4$ and $\gcd(a, r + 1) = 1$. Let X be a surface obtained by a sequence of blow ups as in (1) coming from a divisorial valuation ν with 3 maximal contact values, $\bar{\beta}_0 := a$, $\bar{\beta}_1 := ar^2 - r - 1$ and $\bar{\beta}_2 := \bar{\beta}_0 \bar{\beta}_1 + 1$, and such that the strict transforms of the line at infinity L pass exactly through the first r blown-up points. Notice that $\gcd(\bar{\beta}_0, \bar{\beta}_1) = 1$ and $r \leq \lfloor \bar{\beta}_1 / \bar{\beta}_0 \rfloor$; therefore such a valuation exists.

It can be proved that $\mathrm{Cox}(X)$ is finitely generated (using Theorem 4) and that the anticanonical Iitaka dimension (that is, the Iitaka dimension of $-K_X$) is $-\infty$. Moreover X cannot be obtained with the procedure described in [3, 4] to get surfaces with finitely generated Cox ring. See [10] for complete details.

References

1. M. Artebani, A. Laface, Cox rings of surfaces and anticanonical Iitaka dimension. Adv. Math. **226**, 5252–5267 (2011)
2. C. Birkar, P. Cascini, C.D. Hacon, J. McKernan, Existence of minimal models for varieties of log general type. J. Am. Math. Soc. **23**, 405–468 (2010)
3. A. Campillo, O. Piltant, A. Reguera, Curves and divisors on surfaces associated to plane curves with one place at infinity. Proc. Lond. Math. Soc. **84**, 559–580 (2002)
4. A. Campillo, O. Piltant, A. Reguera, Cones of curves and of line bundles at infinity. J. Algebra **293**, 513–542 (2005)
5. A.M. Castravet, J. Tevelev, Hilbert's 14th problem and cox rings. Compos. Math. **142**, 1479–1498 (2006)
6. D.A. Cox, The homogeneous coordinate ring of a toric variety. J. Algebr. Geom. **4**, 17–50 (1995)
7. L. Ein, R. Lazarsfeld, K. Smith, Uniform approximation of Abhyankar valuation ideals in smooth function fields. Am. J. Math. **125**(2), 409–440 (2003)
8. L. Facchini, V. González-Alonso, M. Lasón, Cox rings of du Val singularities. Le Matematiche **66**, 115–136 (2011)
9. C. Galindo, F. Monserrat, The total coordinate ring of a smooth projective surface. J. Algebra **284**, 91–101 (2005)

10. C. Galindo, F. Monserrat, The cone of curves and the cox ring of rational surfaces given by divisorial valuations. Adv. Math. **290**, 1040–1061 (2016)
11. Y. Hu, S. Keel, Mori dream spaces and GIT. Michigan Math. J. **48**, 331–348 (2000)
12. D. Hwang, J. Park, Cox rings of rational surfaces and redundant blow-ups. Trans. Am. Math. Soc. **368**, 7727–7743 (2016) (To appear)
13. S. Mukai, Counterexample to Hilbert's fourteenth problem for the 3-dimensional additive group, RIMS preprint 1343 (2001)
14. J.C. Ottem, On the cox rings of ^{2} blown up in points on a line. Math. Scand. **109**, 22–30 (2011)
15. D. Testa, A. Várilly-Alvarado, M. Velasco, Big rational surfaces. Math. Ann. **351**, 95–107 (2011)
16. B. Totaro, Hilbert's fourteenth problem over finite fields and a conjecture on the cone of curves. Compos. Math. **144**, 1176–1198 (2008)

Very General Monomial Valuations on $\mathbb{P}^2$ and a Nagata Type Conjecture

Tomasz Szemberg

Abstract The Nagata Conjecture predicts the least degree of a plane curve passing through a set of sufficiently general points with some fixed multiplicity. The purpose of this note is to report on recent new and surprising developments concerning sufficiently general but infinitely near points.

1 History and Motivation

In 1900 David Hilbert announced a list of 23 problems, which he considered important for the development of mathematics in the 20th century. We recall here one of these problems.

Problem 1 (*Hilbert's 14th problem*) Is the ring of invariants of an algebraic group acting on a polynomial ring always finitely generated?

This problem was solved to the negative by Masayoshi Nagata in 1959; see [9]. He studied a graded family of ideals

$$I^{(m)} = I(P_1)^m \cap \cdots \cap I(P_s)^m,$$

where $I(P)$ denotes the saturated ideal of a point $P \in \mathbb{P}^2$ and m runs over positive integers, and showed that for suitable choice of points $P_1, \ldots, P_s$ for every m there

This research has been partially supported by National Science Centre, Poland, grant 2014/15/B/ST1/02197.

T. Szemberg (✉)
Department of Mathematics, Pedagogical University of Cracow,
Podchorążych 2, 30-084 Kraków, Poland
e-mail: tomasz.szemberg@gmail.com

© Springer Nature Switzerland AG 2018
M. Alberich-Carramiñana et al. (eds.), *Extended Abstracts February 2016*,
Trends in Mathematics 9, https://doi.org/10.1007/978-3-030-00027-1_5

exists n such that $(I^{(m)})^n \neq I^{(mn)}$. The last assertion follows from the existence of points $P_1, \ldots, P_s \in \mathbb{P}^2$ such that if a curve C of degree d vanishes at all of these points to order at least m then,

$$d > m\sqrt{s}. \tag{1}$$

Nagata's construction works with $s = r^2$ generic points in $\mathbb{P}^2$, with $r \geq 4$. A great deal of research has been devoted to extending the validity of Nagata's construction for all $s \geq 10$.

Conjecture 2 (Nagata) *The inequality in* (1) *holds for $s \geq 10$ generic points in $\mathbb{P}^2$.*

Around 2000 it has been noticed (see, e.g., [7]) that Nagata's conjecture is closely related to Seshadri constants and hence to the geometry of the nef cone on blow-ups of $\mathbb{P}^2$. In fact, it has been noticed that it is quite irrelevant that the underlying surface is $\mathbb{P}^2$; see, e.g., [1].

Definition 3 (*Seshadri constant*) Let X be a smooth variety and L a big and nef line bundle on X. For a point $P \in X$ the *Seshadri constant of L at P* is the real number

$$\varepsilon(L; P) := \inf_{C \ni P} \frac{(L \cdot C)}{\text{mult}_P(C)} = \sup \left\{ t : f^*L - tE_P \text{ is nef} \right\}.$$

Here, $f : Y \to X$ stands for the blow-up of X at P.

A challenging and completely open problem in the realms of Seshadri constants is the following.

Problem 4 (*(Ir)rationality of Seshadri constants*) Provide an example of X, L, P such that $\varepsilon(L; P)$ is irrational. Or prove that all Seshadri constants are rational.

For our purposes the multi-point version of Definition 3 is more relevant.

Definition 5 (*Multi-point Seshadri constant*) Let X be a smooth variety and L a big and nef line bundle on X. Let $f : Y \to X$ be the blow-up of X at points $P_1, \ldots, P_s$ with exceptional divisors $E_1, \ldots, E_s$, we write $\mathbb{E} = \sum_{i=1}^{s} E_i$. The *multi-point Seshadri constant of L at the set $P_1, \ldots, P_s$* is the real number

$$\varepsilon(L; P_1, \ldots, P_s) := \inf_{C \cap \{P_1, \ldots, P_s\} \neq \emptyset} \frac{(L \cdot C)}{\sum_{i=1}^{s} \text{mult}_{P_i}(C)}.$$

As before, we have equivalently

$$\varepsilon(L; P_1, \ldots, P_s) = \sup \left\{ t : f^*L - t\mathbb{E} \text{ is nef} \right\}.$$

We write $\varepsilon(L; s)$ rather than $\varepsilon(L; P_1, \ldots, P_s)$ if the points $P_1, \ldots, P_s$ are generic. In these terms the Nagata Conjecture simple reads

$$\varepsilon(\mathcal{O}_{\mathbb{P}^2}(1); s) = \frac{1}{\sqrt{s}},$$

for $s \geq 9$. The works of Biran [1] suggest that much more could be true.

Conjecture 6 (Biran–Nagata conjecture) *Let X be a smooth projective surface and let L be a big and nef line bundle on X, Then, $\varepsilon(L; s) = \sqrt{L^2/s}$ for $s \gg 0$ generic points on X.*

See [8] for interesting lower bounds on multi-point Seshadri constants and [3] for a recent overview of various versions and extensions of Nagata's Conjecture.

A somewhat parallel study of linear series of plane curves with imposed fat base points led to the following conjecture formulated independently by Beniamino Segre in 1961, Brian Harbourne in 1986, Alessandro Gimigliano in 1987 and André Hirschowitz in 1988.

Conjecture 7 (SHGH Conjecture) *Let $f : X \to \mathbb{P}^2$ be the blow-up of $\mathbb{P}^2$ in s general points. Let $d, m_1 \geq m_2 \geq \cdots \geq m_s$ be fixed integers with $d \geq m_1 + m_2 + m_3$. Then, the line bundle*

$$M = f^*(\mathcal{O}_{\mathbb{P}^2}(d)) - \sum_{i=1}^{s} m_i E_i$$

is non-special (i.e., the number of global sections of M agrees with a naive count of conditions imposed by fat points).

It is well known that the SHGH Conjecture implies Nagata's Conjecture and that SHGH holds for $s \leq 9$ points; see [2]. Recently, Dumnicki, Küronya, Maclean and the author established an interesting link between the SHGH Conjecture and the rationality problem for Seshadri constants; see [4, Main Th.].

Theorem 8 *Let $s \geq 9$ be an integer for which the SHGH conjecture holds true. Then, either there exist points $P_1, \ldots, P_s \in \mathbb{P}^2$, a line bundle L on $Bl_{P_1, \ldots, P_s}\mathbb{P}^2$ and a point $P \in X$ such that $\varepsilon(L; P)$ is irrational; or the SHGH conjecture fails for $s + 1$ points.*

2 A Valuative Approach to Nagata's Conjecture

Recently Dumnicki, Harbourne, Küronya, Roé and the author studied Nagata's Conjecture from a point of view which allows to make sense of an arbitrary *real* number of points in the conjecture; see [5]. In order to state this result we need to introduce some notation.

Let X be a smooth complex projective surface and let $\mathbb{F}$ be the field of functions on X. Let ν be a rank 1 valuation on $\mathbb{F}$, i.e., the value group of ν is an ordered subgroup of $\mathbb{R}$. For a divisor $D \subset X$, we denote by $\nu(D)$ the value of ν on D (which is computed for the rational function determined by D on an affine chart $U \subset X$). In particular, if ν is centered at a curve $C \subset X$, then $\nu(D)$ is the order of vanishing of D along C. It is also convenient to introduce the following quantity.

$$\mu_D(\nu) = \max \left\{ \nu(D') : \ D' \in |D| \right\}.$$

The subadditivity of μ with respect to the divisor D allows us to work with an asymptotic number

$$\widehat{\mu}_D(\nu) = \lim_{k \to \infty} \frac{\mu_{kD}(\nu)}{k}.$$

Recall that for a divisor D on a surface X its volume is defined as

$$\mathrm{vol}(D) = \lim_{k \to \infty} \frac{h^0(X, kD)}{k^2/2}.$$

We are interested here mainly in $X = \mathbb{P}^2$ and the valuation ν centered at a closed point of $\mathbb{P}^2$. Let $I_m = \{ f \in \mathcal{O}_{\mathbb{P}^2} : \ \nu(f) \geq m \}$. Then, the *volume* of the valuation ν is the real number

$$\mathrm{vol}(\nu) = \lim_{m \to \infty} \frac{\dim_{\mathbb{C}}(\mathcal{O}_{\mathbb{P}^2}/I_m)}{m^2/2}.$$

For the general definition of the volume of a valuation; see [6]. The quantities introduced above are linked in the following way.

Proposition 9 *Let D be a big divisor on a smooth complex projective surface X and let ν be a real valuation centered at a point $P \in X$. Then,*

$$\widehat{\mu}_D(\nu) \geq \sqrt{\mathrm{vol}(D)/\mathrm{vol}(\nu)}. \tag{2}$$

Definition 10 (*A minimal valuation*) For $X = \mathbb{P}^2$ and D a line in $\mathbb{P}^2$ we say that a valuation μ is *minimal* if there is the equality in (2).

Still on $\mathbb{P}^2$ with homogeneous coordinates $[X : Y : Z]$ and in the affine chart $x = X/Z, \ y = Y/Z$ and for the valuation ν centered in $(0, 0)$ we have that $\mu_d(\nu) = \max \{ \nu(f) \mid f \in \mathbb{C}[x, y], \ \deg(f) \leq d \}$ and $\widehat{\mu}(\nu) = \lim_{d \to \infty} \frac{\mu_d(\nu)}{d}$.

Definition 11 (*Quasi-monomial valuation*) Let $\xi(x) \in \mathbb{C}[[x]]$ be a power series with $\xi(0) = 0$. Let $t \geq 1$ be a fixed real number and let θ be an element transcendental over $\mathbb{C}$. Then, $\nu(\xi, t; f) := \mathrm{ord}_0 f(x, \xi(x) + \theta x^t)$ defines a quasi-monomial valuation $\nu(\xi, t; \cdot)$.

The following statement generalizes Nagata's Conjecture; see [5, Conj. 2.4].

Conjecture 12 *For $\xi \in \mathbb{C}[[x]]$ sufficiently general and for all $t \geq 8 + 1/36$ the valuation $\nu_t = \nu(\xi, t; \cdot)$ is minimal, i.e., $\widehat{\mu}(\nu_t) = \sqrt{t}$.*

Some partial support for the above Conjecture is provided by the next Theorem. Let $F_{-1} = 1$, $F_0 = 0$ and $F_{i+1} = F_i + F_{i-1}$ be the Fibonacci sequence, and let $\phi = (1 + \sqrt{5})/2 = \lim F_{i+1}/F_i$ be the "golden ratio".

Theorem 13 *(i) For $t \in [1, \phi^4]$ we have, for all odd $i \geq 1$,*

$$\widehat{\mu}(\nu_t) = \begin{cases} \frac{F_{i-2}}{F_i} t & \text{if } t \in \left[\frac{F_i^2}{F_{i-2}^2}, \frac{F_{i+2}}{F_{i-2}} \right], \\ \frac{F_{i+2}}{F_i} & \text{if } t \in \left[\frac{F_{i+2}}{F_{i-2}}, \frac{F_{i+2}^2}{F_i^2} \right]. \end{cases}$$

(ii) For $t \in [\phi^4, 7 + 1/9]$ we have

$$\widehat{\mu}(\nu_t) = \begin{cases} \frac{1+t}{3} & \text{if } t \in [\phi^4, 7], \\ \frac{8}{3} & \text{if } t \in [7, 7 + 1/9]. \end{cases}$$

In particular, there exists a sequence of rational squares $t < 8$ with $\widehat{\mu}(\nu_t) = \sqrt{t}$ with an accumulation point at ϕ^4.

A serious challenge towards Conjecture 12 is the following.

Problem 14 Establish the equality $\widehat{\mu}(\nu_t) = \sqrt{t}$ for some rational squares > 9.

References

1. P. Biran, Constructing new ample divisors out of old ones. Duke Math. J. **98**, 113–135 (1999)
2. C. Ciliberto, Geometric aspects of polynomial interpolation in more variables and of Waring's problem, in *European Congress of Mathematics, Vol. I (Barcelona, 2000)*. Progress in Mathematics, vol. 201 (Birkhäuser, Basel, 2001), pp. 289–316
3. C. Ciliberto, B. Harbourne, R. Miranda, J. Roé, Variations on Nagata's conjecture. Clay Math. Proc. **18**, 185–203 (2013)
4. M. Dumnicki, A. Küronya, C. Maclean, T. Szemberg, Rationality of Seshadri constants and the Segre–Harbourne–Gimigliano–Hirschowitz conjecture. Adv. Math. **303**, 1162–1170 (2016)
5. M. Dumnicki, B. Harbourne, A. Küronya, J. Roé, T. Szemberg, Very general monomial valuations of $\mathbb{P}^2$ and a Nagata type conjecture. Commun. Anal. Geom. **25**, 125–161 (2017)
6. L. Ein, R. Lazarsfeld, K. Smith, Uniform approximation of Abhyankar valuation ideals in smooth function fields. Am. J. Math. **125**, 409–440 (2003)
7. B. Harbourne, Seshadri constants and very ample divisors on algebraic surfaces. J. Reine Angew. Math. **559**, 115–122 (2003)
8. B. Harbourne, J. Roé, Discrete behavior of Seshadri constants on surfaces. J. Pure Appl. Algebr. **212**, 616–627 (2008)
9. M. Nagata, On the 14-th problem of Hilbert. Am. J. Math. **81**, 766–772 (1959)

Valuations on Equicharacteristic Complete Noetherian Local Domains

Bernard Teissier

Abstract Given an equicharacteristic Noetherian complete local domain R and a rational valuation ν on R we show that there exist an algebra $S = k[\widehat{(u_i)_{i \in I}}]$ equipped with a weight, or monomial valuation, and a surjection $\pi \colon S \to R$ such that the valuation ν is induced by the weight on S in the sense that $\nu(x)$ is the maximum weight of counterimages of x in S. Moreover, the kernel of π is generated by overweight deformations of binomials corresponding to a generating system of the relations between generators $(\gamma_i)_{i \in I}$ of the semigroup $\Gamma = \nu(R \setminus \{0\})$. In all this, the index set I is an ordinal $\leq \omega^{\dim R}$.

1 Introduction

Given an equicharacteristic complete Noetherian local domain R with algebraically closed residue field k, we study the relation between a zero-dimensional valuation ν of R centered at the maximal ideal and its associated graded ring $\mathrm{gr}_\nu R$ with respect to the filtration defined by the valuation. We shall be interested mostly in *rational* valuations, in the sense that the extension $R/m \subset R_\nu/m_\nu$ is trivial, where R_ν is the valuation ring of ν. Then by general properties of valuations, each nonzero homogeneous component of the k-algebra $\mathrm{gr}_\nu R$ is a 1-dimensional k-vector space. In this case the graded algebra is essentially the semigroup algebra $k[t^\Gamma]$ where $\Gamma = \nu(R \setminus \{0\})$ is the semigroup of values of the valuation. The main idea is that in the event that $\mathrm{gr}_\nu R$ is a finitely generated k-algebra, some of the birational toric maps which provide embedded pseudo-resolutions for the affine toric variety corresponding to $\mathrm{gr}_\nu R$ (see [2]) also provide local uniformizations for ν on R, and when $\mathrm{gr}_\nu R$ is not finitely generated, then the same should be true for some of the birational toric maps which

B. Teissier (✉)
Institut de Mathématiques de Jussieu - Paris Rive Gauche, UMR 7586 du CNRS,
Bâtiment Sophie Germain, Case 7012, 75205 Paris Cedex 13, France
e-mail: bernard.teissier@imj-prg.fr

© Springer Nature Switzerland AG 2018
M. Alberich-Carramiñana et al. (eds.), *Extended Abstracts February 2016*,
Trends in Mathematics 9, https://doi.org/10.1007/978-3-030-00027-1_6

(pseudo) resolve the affine variety defined by a well chosen finite subset of the set of binomial equations describing the relations between the generators of Γ. If $\mathrm{gr}_\nu R$ is a finitely generated k-algebra, or equivalently Γ is a finitely generated semigroup, the valuation ν is necessarily Abhyankar (for zero-dimensional valuations this means that the value group is $\mathbf{Z}^r$ with $r = \dim R$). The converse is true up to a birational map on $\mathrm{Spec}\,R$ followed by localization at the center of the valuation and completion; see [4].

In general, the semigroup of a valuation on a Noetherian local domain is well ordered, so that it has a minimal system of generators

$$\Gamma = \langle \gamma_1, \ldots, \gamma_i, \ldots \rangle = \langle (\gamma_i)_{i \in I} \rangle$$

indexed by an ordinal $I \leq \omega^h$, where ω is the ordinal of $\mathbf{N}$ and h is the rank, or (Archimedian) height, of the valuation. If Γ is a numerical semigroup, it is finitely generated by a classical result of Dickson; this is the case when R is one dimensional.

As shown in [3], the fact that R is Noetherian and the valuation ν is rational imply the existence of a presentation of $\mathrm{gr}_\nu R$ as a quotient of a polynomial ring, possibly in countably many variables, by a binomial ideal:

$$k[(U_i)_{i \in I}]/(U^{m^\ell} - \lambda_\ell U^{n^\ell})_{\ell \in L} \simeq \mathrm{gr}_\nu R, \quad \text{with } \lambda_\ell \in k^*.$$

Here the variables U_i are in bijection with the generators γ_i of Γ and the isomorphism sends each U_i to an element $\overline{\xi}_i$, of degree γ_i, of a minimal system of generators of the k-algebra $\mathrm{gr}_\nu R$.

In order to relate the the embedded pseudo-resolutions for the (generalized) affine toric variety corresponding to $\mathrm{gr}_\nu R$ with local uniformizations of the valuation ν on R, the main tools are the concept of *overweight deformation* of a prime binomial ideal, and the *valuative Cohen Theorem*.

2   Weights and Overweight Deformations

For this summary, a weight on $k[(u_i)_{i \in I}]$ or $k[[(u_i)_{i \in I}]]$ will be a morphism of semigroups $w \colon M(I) \to \Phi_{\geq 0}$, where $M(I)$ denotes the semigroup of monomials in the u_i, and Φ a totally ordered abelian group, which attributes to each variable u_i a weight $w(u_i) = \gamma_i \in \Phi_{\geq 0}$. In our case, Φ will be the value group of the valuation and the image of the map w is the semigroup $\Gamma \subset \Phi_{\geq 0}$.

A weight is compatible with a binomial ideal (an ideal generated by binomials) if each generating binomial is homogeneous. The weight of a polynomial, or a series, is the minimum weight of its terms.

Given a weight which is compatible with it, an *overweight deformation* of a binomial is an expression

$$F = u^m - \lambda_{mn} u^n + \sum_{w(u^p) > w(u^m)} c_p u^p \in k[[(u_i)_{i \in I}]].$$

If we have any number of binomials and a compatible weight we do the same and consider the deformations

$$F_\ell = u^{m^\ell} - \lambda_\ell u^{n^\ell} + \sum_{w(u^p) > w(u^{m^\ell})} c_p^{(\ell)} u^p \in k[[(u_i)_{i \in I}]].$$

One has to add the condition that *the initial binomials of the F_ℓ generate the ideal F_0 of initial forms of the elements of the ideal F generated by the F_ℓ.*

Let us consider an overweight deformation (F_ℓ) of a prime binomial ideal F_0 in the power series ring $k[[u_1, \ldots, u_N]]$, and the map

$$\pi \colon k[[u_1, \ldots, u_N]] \to R = k[[u_1, \ldots, u_N]]/(F_1, \ldots, F_s).$$

Proposition 1 *(i) The map which associates to $x \in R$ the maximum of the weights of the elements of $\pi^{-1}(x)$ is well defined and is a valuation ν on R.*

(ii) The associated graded ring of $k[[u_1, \ldots, u_N]]$ with respect to the weight filtration is $k[U_1, \ldots, U_N]$ and the map

$$\Pi \colon k[U_1, \ldots, U_N] \to k[U_1, \ldots, U_N]/F_0 = \mathrm{gr}_\nu R$$

is the associated graded map of π with respect to the weight and valuation filtrations.

(iii) Given $\tilde{x} \in \pi^{-1}(x)$, we have $w(\tilde{x}) = \nu(x)$ if and only if $\mathrm{in}_w(\tilde{x}) \notin F_0$.

3 The Power Series Ring Adapted to Γ

Let $(u_i)_{i \in I}$ be variables indexed by the elements of the minimal system of generators $(\gamma_i)_{i \in I}$ of the semigroup Γ of the rational valuation ν on R. Give each u_i the weight $w(u_i) = \gamma_i$ and let us consider the set of power series $S = \sum_{e \in E} d_e u^e$ where $(u^e)_{e \in E}$ is any set of monomials in the variables u_i and $d_e \in k$.

By a theorem of Campillo–Galindo (see [1]), the semigroup Γ being well ordered is combinatorially finite, which means that for any $\phi \in \Gamma$ the number of different ways of writing ϕ as a sum of elements of Γ is finite. This is equivalent to the fact that the set of exponents e such that $w(u^e) = \phi$ is finite: for any given series the map $w \colon E \to \Gamma$, $e \mapsto w(u^e)$ has finite fibers. Each of these fibers is a finite set of monomials in variables indexed by a totally ordered set, and so can be given the lexicographical order and order-embedded into an interval $1 \le i \le n$ of $\mathbf{N}$. and thus produce an embedding $E \subset (\Gamma \times \mathbf{N})_{lex}$ which induces a total order on E, for which it is well ordered. When E is the set of all monomials, this gives a total monomial order.

The combinatorial finiteness implies that this set of series $S = \sum_{e \in E} d_e u^e$ is a k-algebra, which we denote by $k[\widehat{(u_i)_{i \in I}}]$. It is endowed with a weight $w(S)$, which is the minimum weight of the terms of S, and a topology defined by the weight filtration. It is shown in [4] that the algebra is spherically complete with respect to the monomial valuation given by the weight. Since the weights of the elements of a series form a well ordered set and only a finite number of terms of the series have minimum weight, the associated graded ring of $k[\widehat{(u_i)_{i \in I}}]$ with respect to the filtration by weights is the polynomial ring $k[(U_i)_{i \in I}]$.

Proposition 2 (The valuative Cohen Theorem: see [4]) *Assuming that the local Noetherian equicharacteristic domain R is complete, with a rational valuation ν, and fixing a field of representatives $k \subset R$, there exist choices of representatives $\xi_i \in R$ of the $\overline{\xi_i}$ generating the k-algebra $\mathrm{gr}_\nu R$ such that the surjective map of k-algebras $k[(U_i)_{i \in I}] \to \mathrm{gr}_\nu R$, $U_i \mapsto \overline{\xi_i}$, is the associated graded map of a continuous surjective map*

$$k[\widehat{(u_i)_{i \in I}}] \to R, \ u_i \mapsto \xi_i,$$

of topological k-algebras, with respect to the weight and valuation filtrations respectively. The kernel of this map can be generated up to closure by overweight deformations of binomials generating the kernel of $k[(U_i)_{i \in I}] \to \mathrm{gr}_\nu R$, $U_i \mapsto \overline{\xi_i}$. If ν is of rank one or if Γ is finitely generated, any choice of representatives ξ_i is permitted.

Note that here we have generalized the concept of overweight deformation to the case of countably many binomials.

Corollary 3 *Let ν be a rational valuation on a complete equicharacteristic Noetherian local domain R. If the semigroup Γ of the valuation on R is finitely generated, the ring R is obtained by an overweight deformation from the quotient of a power series ring by the binomial ideal encoding relations between the generators of Γ.*

As a consequence of this, it is a combinatorial problem, relatively easy in the rank one case, to show that if in addition the residue field k is algebraically closed some of the embedded toric resolutions of $\mathrm{gr}_\nu R$ give local uniformization for ν on Spec R, once it is re-embedded in the same ambient space as Spec $\mathrm{gr}_\nu R$. Associated with the fact recalled above that the semigroups of values of rational Abhyankar valuations on excellent equicharacteristic local domains are finitely generated up to ν-modification and completion, this gives a proof of local uniformization for Abhyankar valuations of equicharacteristic excellent local domains with an algebraically closed residue field.

In the general case of a rational valuation ν of a complete equicharacteristic Noetherian local domain the valuative Cohen theorem allows us to produce a sequence of Abhyankar *semivaluations* ν_B of R of, that is, Abhyankar valuations on r-dimensional quotients R/K_B of R, where r is the rational rank of ν, which are indexed by finite subsets B ordered by inclusion of the index set I and have the property that for any element $x \in R$, the valuation $\nu(x)$ is equal to $\nu_B(x)$ for large enough B. This is an extension of the abyssal phenomenon which was described for the plane in [3] and an asymptotic approximation property of rational valuations.

References

1. A. Campillo, C. Galindo, On the graded algebra relative to a valuation. Manuscripta Math. **92**, 173–189 (1997)
2. P. González Pérez, B. Teissier, Embedded resolutions of non necessarily normal affine toric varieties. Comptes-rendus Acad. Sci. Paris, Ser. 1 **334**, 379–382 (2002)
3. B. Teissier, Valuations, deformations, and toric geometry, in *Valuation Theory and Its Applications, Vol. II*. Fields Institute Communications, vol. 33 (AMS, Providence, 2003), pp. 361–459
4. B. Teissier, Overweight deformations of affine toric varieties and local uniformization, in *Valuation Theory in Interaction, Proceedings of the Second International Conference on Valuation Theory, Segovia–El Escorial, 2011*, ed. by A. Campillo, F.-V. Kuhlmann, B. Teissier. Congress Reports Series (European Mathematical Society Publishing House, 2014), pp. 474–565
5. J. Tevelev, On a question of B. Teissier. Collectanea Math. **65**(1), 61–66 (2014). (Published on line February 2013). https://doi.org/10.1007/s13348-013-0080-9

Desingularization by char(X)-Alterations

Michael Temkin

Abstract This is an extended abstract of my talk at the "Workshop on Positivity and Valuations" at Centre de Recerca Matemàtica. The talk was devoted to my recent work (Temkin, Tame distillation and desingularization by p-alterations, [8]), in which I prove that any qe integral Noetherian scheme X can be desingularized by an alteration $X' \to X$ whose degree $[k(X') : k(X)]$ is only divisible by primes non-invertible on X. I had already reported on this work in the Oberwolfach Research Institute for Mathematics, and the first two sections of this abstract are close to my Oberwolfach report. In addition, I added a new third section where further conjectures and directions of research are discussed.

1 The Main Result

1.1 Desingularization

One of the central conjectures of algebraic geometry and adjacent areas is the desingularization conjecture asserting that for any integral algebraic variety X there exists a proper birational morphism $f : X' \to X$ such that the variety X' is regular. In addition, one conjectures that given a closed subset $Z \subsetneq X$ one can arrange that $Z' = f^{-1}(Z)$ is an snc divisor. In fact, there are various stronger forms of the conjecture that cover functoriality of the resolution, etc., but they are not relevant for this work. Also, it was conjectured by Grothendieck and is widely believed that the same desingularization result holds for any quasi-excellent integral scheme X.

The desingularization conjecture was proved in characteristic zero by Hironaka for schemes of finite type over a local quasi-excellent ring (see [5]) and it was

M. Temkin (✉)
Einstein Institute of Mathematics, The Hebrew University of Jerusalem,
Giv'at Ram, 91904 Jerusalem, Israel
e-mail: temkin@math.huji.ac.il

© Springer Nature Switzerland AG 2018
M. Alberich-Carramiñana et al. (eds.), *Extended Abstracts February 2016*,
Trends in Mathematics 9, https://doi.org/10.1007/978-3-030-00027-1_7

proved for all quasi-excellent schemes over $\mathbf{Q}$ by Temkin [7]. Also, the conjecture was established very recently for quasi-excellent threefolds by Cossart–Piltant [1]. Already for varieties of positive characteristic the conjecture is widely open and very difficult in dimensions starting with 4.

1.2 de Jong's Altered Desingularization

de Jong found a very successful weakening of the desingularization conjecture: its proof is relatively simple (e.g., when comparing with [1, 5]), and yet, it has numerous applications. Namely, de Jong proved in [2, Thm. 4.1] that for any integral scheme X of finite type over a quasi-excellent base of dimension 2 (using [1] this can be pushed to dimension 3) there exists an *alteration* $f : X' \to X$, i.e., a proper dominant generically finite morphism between integral schemes, such that X' is regular. In addition, if $Z \subsetneq X$ is closed then one can arrange that $Z' = f^{-1}(Z)$ is an snc divisor.

1.3 Gabber's l'-Altered Desingularization

One can deduce from de Jong's theorem various cohomological applications that before de Jong's work were only known to be consequences of the desingularization conjecture. However, usually de Jong's theorem imposes one essential restriction on these applications: the coefficients of the cohomology theory should contain $\mathbf{Q}$. In order to deal with cohomology theories where a prime l is not inverted, e.g. $\mathbf{Z}/l\mathbf{Z}$ or $\mathbf{Z}_l$-cohomology, Gabber strengthened de Jong's theorem as follows: keep the assumptions of de Jong's theorem and assume that l is a prime number invertible on X, then the desingularizing alteration $f : X' \to X$ can be chosen so that l does not divide the degree $\deg(f) = [k(X') : k(X)]$; see [6, Thm. 2.1]. Such alterations are called l'-alterations.

1.4 char(X)-Altered Desingularization

It is a natural question if Gabber's theorem can be strengthened so that $\deg(f)$ is not divisible by two (or more) fixed primes invertible on X. In my recent work [8] I answer this affirmatively, in fact, I prove that one can avoid all invertible primes simultaneously. By a char(X)-alteration we mean an alteration $X' \to X$ whose degree is only divisible by primes non-invertible on X. The main result of [8] is that if X is of finite type over a quasi-excellent threefold and $Z \subsetneq X$ is closed then there exists a char(X)-alteration $f : X' \to X$ such that X' is regular and $f^{-1}(Z)$ is an snc divisor.

In particular, if X is of characteristic zero then f is a desingularization, and if X is of characteristic p then $\deg(f) = p^n$.

2   The Method

2.1 l'-Altered Desingularization

de Jong refined his theorem in [3] as follows: the altered desingularization $f: X' \to X$ can be chosen so that the alteration $g: X'/\mathrm{Aut}_X(X') \to X$ is generically radicial (in particular, $\deg(g) = p^n$ where p is the characteristic exponent of $k(X)$). Gabber observed that the l-Sylow subgroup G_l of $G = \mathrm{Aut}_X(X')$ acts tamely on X' whenever l is invertible on X and proved a general difficult theorem on tame actions implying that there exists a G_l-equivariant modification $X'' \to X'$ such that $Y = X''/G_l$ is regular. In particular, $Y \to X$ is an l'-altered desingularization of X.

2.2   Tame Distillation

Note that if there exists a subgroup $H \subseteq G$ acting tamely on X' and $|G/H|$ is only divisible by primes non-invertible on X then the same argument as above works with G_l replaced by H. In general, such an H does not have to exist and the main new tool of [8] is the following result that asserts that such an H exists if one enlarges the alteration $X' \to X$.

Tame distillation theorem, see [8, Thm. 3.3.6]: for any alteration $X' \to X$ of quasi-excellent schemes there exists an alteration $Y' \to X'$ such that the composition $Y' \to X$ factors into a composition of a tame Galois covering $Y' \to Y$ and a char(X)-alteration $Y \to X$.

2.3   char(X)-Altered Desingularization

The tame distillation does not apply directly to Gabber's argument. Indeed, in order to construct a large enough tamely acting group H we have to replace the regular scheme X' with its alteration Y', and one cannot ensure that Y' is also regular. However, Illusie and Temkin discovered in [6, Sect. 3] a more flexible proof of Gabber's theorem which is also based on division by l-Sylow subgroups (the main motivation for finding that proof was to extend Gabber's theorem to morphisms of finite type; see [6, Thm. 3.5]). Once one replaces l-Sylow subgroups by the subgroups provided by the tame distillation theorem, the argument of Illusie–Temkin applies almost verbatim and yields a proof of the char(X)-alteration theorem. We refer to [8, Thm. 4.3.1] and its proof for details.

3   Future Research

In the last section, let us discuss conjectures that are not covered by [8] but, as the author expects, can be established by similar methods and ideas.

3.1   *char*(X)-*Altered Local Uniformization*

In strong enough desingularization theorems one resolves X by a modification $f: X' \to X$ which only modifies the singular locus of X, i.e., f is an isomorphism over the regular locus X_{reg} of X. It is a natural question if altered desingularization has an analogous strengthening. Since we use alterations, one cannot expect f to be an isomorphism over X_{reg} and the best one can hope for is to achieve that f is étale over X_{reg}. Moreover, it can freely happen that X_{reg} has no non-trivial finite étale coverings, hence we should allow desingularizing morphisms f which are étale but not finite over X_{reg}. So, it is natural to consider any f which is a covering for the *alteration topology* τ generated by alterations and Zariski coverings. I expect that the following conjecture can be proved by using the ideas from [8]:

Conjecture 1 *For any integral qe scheme X with a closed subset $Z \subsetneq X$ there exists a morphism $f: X' \to X$ étale over $X_{\text{reg}} \setminus Z$ and such that X' is regular, $Z' = f^{-1}(Z)$ is an snc divisor, f is a covering for the alteration topology, and for any $x \in X_{\text{reg}} \setminus Z$ and $x' \in f^{-1}(x)$, the degree $[k(x') : k(x)]$ is only divisible by primes non-invertible on X.*

3.2   *char*(X)-*Altered Local Semistable Reduction*

Assume now that R is a valuation ring with field of fractions K. Set $S = \text{Spec}(R)$ and $\eta = \text{Spec}(K)$, and assume that X is an integral flat S-scheme of finite type with smooth generic fiber X_η. By the above conjecture, one can find a resolution $f: X' \to X$ such that f is a τ-covering étale over X_η. However, the morphism $X' \to S$ can be very far from being smooth since it may have non-reduced fibers. As in the classical semistable reduction conjecture, it is natural to expect that the situation improves once one allows to extend R. Given a finite extension K'/K with a valuation ring R' of K' dominating R, we will use the notation $S' = \text{Spec}(R')$ and $\eta' = \text{Spec}(K')$. The following conjecture is a local altered version of the semistable reduction conjecture, that, as I expect, can be proved using the circle of ideas of [8]:

Conjecture 2 *Let R be a valuation ring of residual characteristic exponent p, $K = \text{Frac}(R)$, $S = \text{Spec}(R)$, $\eta = \text{Spec}(K)$. Assume that X is an integral flat S-scheme of finite type such that X_η is smooth. Then there exists an extension K'/K of degree p^n with a valuation ring R' of K' dominating R such that the normalization*

$X' = \text{Nor}(X \times_S S')$ *possesses a covering* $f : Y' \to X'$ *in the topology of alterations satisfying the following conditions:* Y' *is semistable over* S', f *is étale over* $X'_{\eta'}$ *and for any point* $y \in Y'_{\eta'}$ *with* $x = f(y)$ *the degree* $[k(y) : k(x)]$ *is a power of* p.

If R is real-valued complete then there is also an analogue of the Conjecture 2, where schemes and spectra are replaced with formal schemes and formal spectra. This formal version implies the following conjecture on the local structure of non-Archimedean analytic spaces.

Conjecture 3 *Assume that* k *is a complete real-valued field of residual characteristic exponent* p. *Then for any rig-smooth analytic non-Archimedean space* X *there exist an extension* l/k *of degree* p^n *and étale morphisms* $f_i : Y_i \to X$ *such that the following conditions are satisfied: each* Y_i *is an* l*-affinoid space* $\mathcal{M}(\mathcal{A}_i)$ *whose maximal affine formal model* $\text{Spf}(\mathcal{A}_i^\circ)$ *is semistable over* $\text{Spf}(l^\circ)$, *the analytic domains* $f_i(Y_i)$ *form an admissible covering of* X *(in other words,* $Y = \coprod_i Y_i \to X$ *is a covering for the Tate-étale topology) and, for any* $y \in Y_i$ *with* $x = f_i(y)$, *the degree* $[\mathcal{H}(y) : \mathcal{H}(x)]$ *is a power of* p.

Note that the weaker version of this conjecture, where no restriction on the degrees of l/k and $\mathcal{H}(y)/\mathcal{H}(x)$ is imposed, is a theorem of Hartl [4]. The relation between the conjecture and the theorem of Hartl is the same as the relation between the char(X)-altered desingularization theorem of [8] and the alteration theorem of de Jong.

References

1. V. Cossart, O. Piltant, Resolution of singularities of arithmetical threefolds II, Preprint, arXiv:1412.0868
2. J. de Jong, Smoothness, semistability and alterations. Inst. Hautes études Sci. Publ. Math. **83**, 51–93 (1996)
3. J. de Jong, Families of curves and alterations. Ann. Inst. Fourier **47**, 599–621 (1997)
4. U. Hartl, Semi-stable models for rigid-analytic spaces. Manuscripta Math. **110**, 365–380 (2003)
5. H. Hironaka, Resolution of singularities of an algebraic variety over a field of characteristic zero. Ann. Math. **79**, 109–326 (1964)
6. L. Illusie, M. Temkin, Exposé X. Gabber's modification theorem (log smooth case). Astérisque **363–364**, 167–212 (2014)
7. M. Temkin, Desingularization of quasi-excellent schemes in characteristic zero. Adv. Math. **219**, 488–522 (2008)
8. M. Temkin, Tame distillation and desingularization by p-alterations, Ann. Math. **186**, 97–126 (2017)

Semigroup and Poincaré Series for Divisorial Valuations

Willem Veys

Abstract Let V be a finite set of divisorial valuations coming from a modification of K^d, where K is a field. We present results on the semigroup of values and the Poincaré series associated to V, assuming that V has a finite generating sequence. First, if K is infinite, this semigroup is finitely generated. Secondly, for any K, the Poincaré series associated to V is a rational function whose denominator can be expressed in terms of the valuation vectors of the elements in the generating sequence.

1 Introduction

The semigroup of values and the Poincaré series are introduced to study properties of a finite set of divisorial valuations. We mention for example [5–8, 10]. The two-dimensional case is the most understood and well behaved. There it is known (see [7]) that a set of valuations coming from a modification of K^2, with K algebraically closed, has a finite generating sequence and that the semigroup of values is finitely generated. Besides, there exists an explicit description of the generating sequence in terms of the dual graph of the modification; this graph even determines the generators of the semigroup. Furthermore, when $K = \mathbb{C}$, there exists an explicit formula for the Poincaré series in terms of the topology of the exceptional locus of the modification.

In higher dimensions, Lemahieu [10] looks at toric constellations on $\mathbb{C}^d$. In particular, all valuations that are considered, are monomial. She shows that then the Poincaré series equals $1/\prod_{i=1}^{d}(1 - \underline{t}^{\nu(x_i)})$, where $x_1, \ldots, x_d$ is the coordinate system of $\mathbb{C}^d$ and where $\nu(x_i)$ is the valuation vector of x_i.

Here, we present in arbitrary dimension the work of Van Langenhoven–Veys [13], generalizing the results above, assuming that the given set of valuations has a finite generating sequence.

We use the notation $\mathbb{N} = \{n \in \mathbb{Z} \mid n \geqslant 0\}$ and $\mathbb{N}^* = \{n \in \mathbb{Z} \mid n > 0\}$. The homogeneous maximal ideal of the ring $K[x_1, \ldots, x_d]$, i.e., $(x_1, \ldots, x_d)$, is denoted by $\mathfrak{m}$.

W. Veys (✉)
KU Leuven, Department of Mathematics, Celestijnenlaan 200B, 3001 Leuven, Belgium
e-mail: wim.veys@kuleuven.be

© Springer Nature Switzerland AG 2018
M. Alberich-Carramiñana et al. (eds.), *Extended Abstracts February 2016*,
Trends in Mathematics 9, https://doi.org/10.1007/978-3-030-00027-1_8

2 Preliminaries

Definition 1 Let K be a field. In this text a *modification* π of K^d is a composition of blow-ups

$$X_r \xrightarrow{\pi_r} X_{r-1} \xrightarrow{\pi_{r-1}} \cdots \xrightarrow{\pi_1} X_0 \xrightarrow{\pi_0} K^d,$$

$\pi = \pi_0 \circ \pi_1 \circ \cdots \circ \pi_r$, where π_0 is the blow-up of K^d at the origin. All the other π_σ, $1 \leq \sigma \leq r$, are blow-ups at smooth, irreducible centers $Z_\sigma (\subseteq X_{\sigma-1})$ that are defined over K, have codimension at least 2, and are contained in and have normal crossings with the exceptional locus of $\pi_0 \circ \cdots \circ \pi_{\sigma-1}$.

Denote the exceptional locus of π_σ, as well as its consecutive strict transforms, by E_σ. For a polynomial $g \in K[x_1, \ldots, x_d] \setminus \{0\}$, let $\nu_\sigma(g)$ be the vanishing order of $g \circ \pi$ along E_σ (and $\nu_\sigma(0) = \infty$). The map ν_σ defines a divisorial valuation on $K(x_1, \ldots, x_d)$.

Now let us fix s different components $E_{\sigma_1}, \ldots, E_{\sigma_s}$ of the exceptional divisor $E = \pi^{-1}(0)$, let $\nu_i = \nu_{\sigma_i}$ for $i \in \{1, \ldots, s\}$, and take $V = \{\nu_1, \ldots, \nu_s\}$. When $g \in K[x_1, \ldots, x_d]$, we call $\nu(g) = (\nu_1(g), \ldots, \nu_s(g))$ the valuation vector of g.

Remark 2 When K is a field of characteristic 0, any finite set of divisorial valuations centered at the origin of K^d can be viewed as such a set V (using resolution of indeterminacies). So, in fact our setup is very general.

Definition 3 The *semigroup of values* of V is the additive subsemigroup of $\mathbb{N}^s$ given by

$$S_V = \{\nu(g) = (\nu_1(g), \ldots, \nu_s(g)) \mid g \in K[x_1, \ldots x_d] \setminus \{0\}\}.$$

Its *saturation* is $S_V^{sat} = \{x \in \mathbb{Z}^s \mid kx \in S_V \text{ for some } k \in \mathbb{N}^*\}$.

Definition 4 Let $V = \{\nu_1, \ldots, \nu_s\}$ be defined as above. Then, for every $v \in \mathbb{Z}^s$, we define its *valuation ideal* $J(v) = \{g \in K[x_1, \ldots, x_d] \mid \nu(g) \geqslant v\}$. Following [3, 6], we define the *Poincaré series* associated to V as

$$P_V(t_1, \ldots, t_s) = \frac{\prod_{i=1}^s (t_i - 1)}{t_1 \cdots t_s - 1} \sum_{v \in \mathbb{Z}^s} d(v) \underline{t}^v,$$

where $d(v) = \dim_K J(v)/J(v + (1, 1, \ldots, 1))$.

Definition 5 Let $\Lambda = \{q_\alpha\}_{\alpha \in A}$ be a subset of $\mathfrak{m}$. A *monomial in* Λ is defined as a finite product in the elements of Λ, namely, $\prod q_\alpha^{M_\alpha}$, with every $M_\alpha \in \mathbb{N}^*$. We call Λ a *generating sequence* for V if for every $v \in \mathbb{N}^s$ the ideal $J(v)$ is generated (as an ideal) by all monomials in Λ that are also in $J(v)$.

3 Results

Theorem 6 *Let K be an infinite (or big enough) field. If $\Lambda = \{q_1, \dots, q_k\}$ is a finite generating sequence for V, then S_V is a finitely generated semigroup.*

In particular, this applies to any toric constellation, since then each V has $\Lambda = \{x_1, \dots, x_d\}$ as a generating sequence.

The proof we provide, gives us a method to compute the generators of S_V. However, this method introduces a lot of extra variables. On the other hand, (generators of) S_V^{sat} can be computed more easily. In our algorithms, hyperplanes defining relevant cones can be computed by Fourier–Motzkin elimination [12], and generators using Normaliz [1].

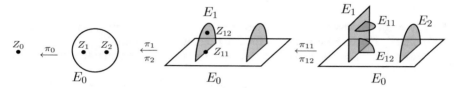

Example 7 We present a set of valuations V, coming from an 'easy' toric constellation on $\mathbb{C}^3$, but with non-saturated semigroup S_V.

We start by the blow-up $\pi_0 \colon X_0 \to \mathbb{C}^3$ at the origin Z_0 with exceptional divisor E_0. This blow-up can be described by three affine charts, say K_1, K_2 and K_3. Next we blow up at the origin Z_1 of K_1, which gives us an extra exceptional divisor E_1 and charts $K_{1.1}$, $K_{1.2}$ and $K_{1.3}$. Then we blow up at the origin Z_2 of K_2, which gives us E_2 and three extra affine charts. After this, we blow up at the origins of $K_{1.1}$ and $K_{1.2}$; this gives rise to the exceptional divisors E_{11} and E_{12}.

It gives us five divisorial valuations $V = \{\nu_0, \nu_1, \nu_2, \nu_{11}, \nu_{12}\}$. Since they are all monomial, $\Lambda = \{x, y, z\}$ is a generating sequence for V. The associated valuation vectors of x, y and z are $(1, 1, 2, 1, 2)$, $(1, 2, 1, 3, 3)$ and $(1, 2, 2, 3, 4)$, respectively.

Although the modification is relatively simple, the semigroup S_V is more complex than one might expect. For one, it is not saturated. Thanks to Bruns–Ichim [1], we can compute the 1820 generators of a certain rational cone used in the proof of Theorem 6. From this we computed the following 22 generators of S_V:

$$(1, 1, 1, 1, 2)\ (2, 3, 2, 3, 6)\ (2, 4, 4, 5, 7)\quad (3, 6, 6, 7, 10)$$
$$(1, 1, 2, 1, 2)\ (2, 3, 2, 4, 6)\ (3, 6, 4, 6, 12)\ (3, 6, 6, 8, 10)$$
$$(1, 2, 1, 2, 3)\ (2, 4, 3, 4, 8)\ (3, 6, 4, 7, 12)\ (4, 8, 8, 12, 15)$$
$$(1, 2, 1, 3, 3)\ (2, 4, 3, 5, 8)\ (3, 6, 4, 8, 12)\ (5, 10, 10, 15, 18)$$
$$(1, 2, 2, 2, 4)\ (2, 4, 3, 6, 8)\ (3, 6, 4, 9, 12)$$
$$(1, 2, 2, 3, 4)\ (2, 4, 4, 4, 7)\ (3, 6, 6, 6, 10)$$

We see that $(6, 12, 12, 18, 22) = \nu(z^6 + x^3 y^4 z) = (1, 2, 2, 3, 4) + (5, 10, 10, 15, 18) \in S_V$. But it turns out that $(3, 6, 6, 9, 11)$ cannot be an element of S_V, and hence S_V is not saturated.

Our method to compute S_V^{sat} yields that $(n_0, n_1, n_2, n_{11}, n_{12}) \in \mathbb{Z}^5$ is an element of S_V^{sat} if and only if it satisfies all the inequalities

$$
\begin{array}{lll}
n_2 - n_0 \geqslant 0 & \left| \begin{array}{l} 2n_0 - n_1 \geqslant 0 \\ 2n_0 - n_2 \geqslant 0 \\ 2n_1 - n_{12} \geqslant 0 \\ 2n_1 - n_0 - n_{11} \geqslant 0 \end{array} \right. & \left| \begin{array}{l} 3n_2 - n_{12} \geqslant 0 \\ 3n_{12} - 4n_1 - n_2 \geqslant 0 \\ 5n_{12} - 3n_2 - 4n_{11} \geqslant 0 \end{array} \right.
\end{array}
$$

It turns out that S_V^{sat} also has 22 generators, more precisely those for S_V above, where $(4, 8, 8, 12, 15)$ is replaced by $(3, 6, 6, 9, 11)$.

Remark 8 In dimensional two the semigroup of values of a modification over an algebraically closed field is always finitely generated and there always exists a finite generating sequence that can be expressed in terms of the dual graph of the modification; see [7].

In higher dimensions, the semigroup of values of a modification is not always finitely generated. Note that in such cases there cannot be a finite generating sequence. The fact that there is different behaviour in higher dimensions, could be expected from examples in similar settings such as [2, Rem. 1.24] and [4].

Our counterexample to confirm that statement is constructed as follows. We start with the blow-up at the origin of $\mathbb{C}^3$ and then we blow up at nine very general points on the first exceptional divisor. This gives us a set V of ten valuations, such that S_V is not finitely generated. Our proof uses the fact that the cone of curves of the blow-up of $\mathbb{P}^2$ at nine very general points is not finitely generated; see [9, II, Lem. 4.12] and [11]. In fact, we construct more precisely an explicit list of generators of S_V^{sat}.

Theorem 9 *If $\Lambda = \{q_1, \ldots, q_k\}$ is a finite generating sequence for V, then the Poincaré series associated to V is a rational function whose denominator equals $\prod_{i=1}^{k}(1 - \underline{t}^{\nu(q_i)})$.*

Our long and technical proof uses the following strategy. We endow the polynomial ring $K[Q_1, \ldots, Q_k]$ with $r + 1$ new *monomial* valuations $\hat{v}_i$, defined by $\hat{v}_i(Q_j) = \nu_i(q_j)$, to which we associate a certain 'mixed' semigroup Σ_V. We also consider a morphism $\phi \colon K[Q_1, \ldots, Q_k] \to K[x_1, \ldots, x_d]$, sending Q_j to the lowest degree homogeneous component of q_j. Then $\ker \phi$ determines a certain ideal of the semigroup Σ_V, whose (finitely many) generators describe in a combinatorial way the dimensions in the definition of the Poincaré series. (Remember that every $\nu(q_i) \geqslant \underline{1}$, because $\Lambda \subseteq \mathfrak{m}$.)

Remark 10 If $\ker \phi = \langle B \rangle$ is generated by one element, then the Poincaré series has the simple expression

$$
P_V(\underline{t}) = \frac{1 - \underline{t}^{\hat{v}(B)}}{\prod_{i=1}^{k}(1 - \underline{t}^{\nu(q_i)})}.
$$

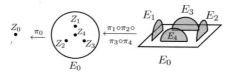

Example 11 Let $\pi_0: X_0 \to \mathbb{C}^3$ be the blow-up at the origin with exceptional divisor E_0. We blow up further at four points in this E_0, say $(1:0:0)$, $(0:1:0)$, $(0:0:1)$ and $(1:1:1)$. This gives us five divisorial valuations, say $\nu_0, \nu_1, \nu_2, \nu_3$ and ν_4.

A somewhat lengthy computation yields that $\Lambda = \{x, y, z, x - z, y - z, x - y\}$ is a generating sequence for the set of valuations $V = \{\nu_0, \nu_1, \nu_2, \nu_3, \nu_4\}$. The corresponding valuation vectors are $\nu(x) = (1, 1, 2, 2, 1)$, $\nu(y) = (1, 2, 1, 2, 1)$, $\nu(z) = (1, 2, 2, 1, 1)$, $\nu(x - z) = (1, 1, 2, 1, 2)$, $\nu(y - z) = (1, 2, 1, 1, 2)$ and $\nu(x - y) = (1, 1, 1, 2, 2)$.

Firstly, we construct the surjective morphism $\phi: \mathbb{C}[Q_1, \ldots, Q_6] \to \mathbb{C}[x, y, z]$, $Q_1 \mapsto x$, $Q_2 \mapsto y$, $Q_3 \mapsto z$, $Q_4 \mapsto x - z$, $Q_5 \mapsto y - z$, $Q_6 \mapsto x - y$. Secondly we define the induced monomial valuations on $\mathbb{C}[Q]$ where the valuations of the Q_i are

	Q_1	Q_2	Q_3	Q_4	Q_5	Q_6
$\hat{v}_0$	1	1	1	1	1	1
$\hat{v}_1$	1	2	2	1	2	1
$\hat{v}_2$	2	1	2	2	1	1
$\hat{v}_3$	2	2	1	1	1	2
$\hat{v}_4$	1	1	1	2	2	2

In this case $\ker \phi = (Q_1 - Q_2 - Q_6, Q_2 - Q_3 - Q_5, Q_4 - Q_5 - Q_6)$. It is pretty complicated to find a basis of the relevant ideal of Σ_V. Knowing such a basis, we can compute that the Poincaré series $P_V(\underline{t})$ is equal to

$$P_V(\underline{t}) = \frac{\begin{aligned}1 + \underline{t}^{(1,1,1,1,1)} - \underline{t}^{(1,2,1,1,1)} - \underline{t}^{(1,1,2,1,1)} - \underline{t}^{(1,1,1,2,1)} - \underline{t}^{(1,1,1,1,2)} \\ + \underline{t}^{(2,3,3,3,2)} + \underline{t}^{(2,3,3,2,3)} + \underline{t}^{(2,3,2,3,3)} + \underline{t}^{(2,2,3,3,3)} - \underline{t}^{(2,3,3,3,3)} - \underline{t}^{(3,4,4,4,4)}\end{aligned}}{\begin{aligned}(1 - \underline{t}^{(1,2,2,1,1)})(1 - \underline{t}^{(1,2,1,2,1)})(1 - \underline{t}^{(1,2,1,1,2)}) \\ (1 - \underline{t}^{(1,1,2,2,1)})(1 - \underline{t}^{(1,1,2,1,2)})(1 - \underline{t}^{(1,1,1,2,2)})\end{aligned}}.$$

Note that this numerator cannot be factored. This is different from the case where π is a modification of $\mathbb{C}^2$; see, for example, [6, 7].

Question 12 *For a given set of divisorial valuations V, how can one 'guess' a generating sequence, and prove efficiently that it is really a generating sequence?*

Question 13 *Is the Poincaré series $P_V(\underline{t})$ for the example in Remark 8 a rational function or not?*

References

1. W. Bruns, B. Ichim, Normaliz: algorithms for affine monoids and rational cones. J. Algebr. **324**(5), 1098–1113 (2010)
2. A. Campillo, G. González-Sprinberg, M. Lejeune-Jalabert, Clusters of infinitely near points. Math. Ann. **306**(1), 169–194 (1996)
3. A. Campillo, F. Delgado, S.M. Gusein-Zade, The Alexander polynomial of a plane curve singularity via the ring of functions on it. Duke Math. J. **117**(1), 125–156 (2003)
4. V. Cossart, C. Galindo, O. Piltant, Un exemple effectif de gradué non noethérien associé à une valuation divisorielle. Ann. Inst. Fourier (Grenoble) **50**(1), 105–112 (2000)
5. S.D. Cutkosky, J. Herzog, A. Reguera, Poincaré series of resolutions of surface singularities. Trans. Am. Math. Soc. **356**(5), 1833–1874 (electronic) (2004)
6. F. Delgado, S.M. Gusein-Zade, Poincaré series for several plane divisorial valuations. Proc. Edinb. Math. Soc. (2) **46**(2), 501–509 (2003)
7. F. Delgado, C. Galindo, A. Núñez, Generating sequences and Poincaré series for a finite set of plane divisorial valuations. Adv. Math. **219**(5), 1632–1655 (2008)
8. C. Galindo, F. Monserrat, Finite families of plane valuations: value semigroup, graded algebra and Poincaré series, in *Zeta Functions in Algebra and Geometry*. Contemporary Mathematics, vol. 566 (American Mathematical Society, Providence, 2012), pp. 189–212
9. J. Kollár, in *Rational Curves on Algebraic Varieties*. Ergebnisse der Mathematik und ihrer Grenzgebiete. 3. Folge. A Series of Modern Surveys in Mathematics, vol. 32 (Springer, Berlin, 1996). [Results in Mathematics and Related Areas. 3rd Series. A Series of Modern Surveys in Mathematics]
10. A. Lemahieu, Poincaré series of a toric variety. J. Algebr. **315**(2), 683–697 (2007)
11. M. Nagata, On rational surfaces. II. Mem. Coll. Sci. Univ. Kyoto Ser. A Math. **33**, 271–293 (1960/1961)
12. A. Schrijver, in *Theory of Linear and Integer Programming*. Wiley-Interscience Series in Discrete Mathematics (Wiley, Chichester, 1986)
13. L. Van Langenhoven, W. Veys, Semigroup and Poincaré series for a finite set of divisorial valuations. Rev. Mat. Complut. **28**(1), 191–225 (2015)

Computing Multiplier Ideals in Smooth Surfaces

Guillem Blanco and Ferran Dachs Cadefau

Abstract We present an algorithm to compute the jumping numbers and the multiplier ideals associated to a given ideal in a regular two-dimensional local ring. All computations are effective in the sense that both the input and the output are ideals generated by equations.

1 Introduction

Multiplier ideals and their associated jumping numbers have proven to be a powerful tool to understand the geometry of singularities. They are defined using a log-resolution of the pair $(X, \mathfrak{a})$ composed by a complex variety X and an ideal $\mathfrak{a} \subseteq \mathcal{O}_{X,O}$. In fact, smaller or more dense jumping numbers can be thought to correspond to "worse" singularities. In [3], Alberich-Carramiñana–Àlvarez-Montaner–Dachs-Cadefau describe a method to find the jumping numbers and multiplier ideals corresponding to an ideal. That method requires the log-resolution of the ideal and ouputs the multiplier ideals by means of a divisor. However, thanks to the results of Alberich-Carramiñana–Àlvarez-Montaner–Blanco [1, 2], we are able to present an algorithm that, starting with a set of generators of the ideal, gives a set of generators of the multiplier ideals.

The first part of this paper is devoted to introduce the main tools required for the algorithm, which is explained in the second part. In the last part, we present an example of the computations.

G. Blanco (✉)
Departament de Matemàtiques, Universitat Politècnica de Catalunya, Av. Diagonal 647,
08028 Barcelona, Spain
e-mail: gblanco92@gmail.com

F. D. Cadefau
Institut für Mathematik, Martin–Luther–Universität Halle-Wittenberg, 06099 Saale, Halle,
Germany
e-mail: ferran.dachscadefau@gmail.com

M. Alberich-Carramiñana et al. (eds.), *Extended Abstracts February 2016*,
Trends in Mathematics 9, https://doi.org/10.1007/978-3-030-00027-1_9

2 Preliminaries and Results

We introduce the main definitions and tools that we will use to present the algorithm. This section is based on the papers [1–3].

2.1 Computing Log-Resolution of Ideals

Let (X, O) be a germ of smooth complex surface and $\mathcal{O}_{X,O}$ the ring of germs of holomorphic functions in a neighborhood of O, that we identify with $\mathbb{C}\{x, y\}$ by taking local coordinates. We also denote $\mathfrak{m} = \mathfrak{m}_{X,O} \subseteq \mathcal{O}_{X,O}$ the maximal ideal. Let $\mathfrak{a} \subseteq \mathcal{O}_X$ be an ideal sheaf. A *log-resolution* of the pair $(X, \mathfrak{a})$, or a log-resolution of $\mathfrak{a}$ for short, is a proper birational morphism $\pi \colon X' \to X$ such that X' is smooth, the preimage of $\mathfrak{a}$ is locally principal, that is $\mathfrak{a} \cdot \mathcal{O}_{X'} = \mathcal{O}_{X'}(-F)$ for some effective Cartier divisor F, and $F + E$ is a divisor with simple normal crossings support where $E = Exc(\pi)$ is the exceptional locus.

From now on, if no confusion arises, we will indistinctly denote by $\mathfrak{a}$ the sheaf ideal or its stalk at O. In this later case we will be considering an ideal $\mathfrak{a} \subseteq \mathcal{O}_{X,O}$.

Any log-resolution of an ideal is a composition of blowing-ups of points infinitely near to O. Hence, attached to $\mathfrak{a}$, there is a pair $\mathcal{K} = (K, v)$ where K is the set of infinitely near points that have been blown-up to reach a minimal log-resolution of $\mathfrak{a}$, and $v \colon K \longrightarrow \mathbb{Z}$ is a valuation map that encodes the coefficients of the exceptional components in F. More precisely, the divisor F decomposes into its affine and exceptional part $F = F_{\mathrm{aff}} + F_{\mathrm{exc}}$ according to its support. If E_i is the exceptional divisor that arises from the blowing-up of a point $p_i \in K$, we have $F_{exc} = \sum_i d_i E_i$ where $v(p_i) = d_i$.

If $\mathfrak{a} = (f)$ is a principal ideal with $f \in \mathcal{O}_{X,O}$, the minimal log-resolution of $\mathfrak{a}$ equals the minimal log-resolution of the reduced curve ξ_{red} of $\xi \colon f = 0$, that is, the composition of blowing-ups of all infinitely near singular points of ξ_{red}. If the ideal $\mathfrak{a} = (a_1, \ldots, a_r) \subseteq \mathcal{O}_{X,O}$ is not principal, the minimal log-resolution π of $\mathfrak{a}$ is no longer straightforwardly deduced from the minimal log-resolutions $\pi_i \colon X'_i \longrightarrow X$ of the principal ideals $\mathfrak{a}_i = (a_i)$ corresponding to each generator. Neither π dominates any π_i, nor the minimal proper birational morphism $\pi' \colon Y \longrightarrow X$ dominating all π_i, which is the minimal log-resolution of the principal ideal $(a_1 \cdots a_r)$, dominates π.

In [1], Alberich-Carramiñana–Àlvarez-Montaner–Blanco describe an algorithm that computes the minimal log-resolution of an ideal $\mathfrak{a} = (a_1, \ldots, a_r) \subseteq \mathcal{O}_{X,O}$ from the minimal log-resolution of $\xi_{\mathrm{red}} \colon a_1 \cdots a_r = 0$ and the minimal log-resolution of each generator a_i. Computing the log-resolution of reduced curves is an already well-known procedure see, for instance [4], therefore minimal log-resolution of ideals can be effectively computed.

Any ideal $\mathfrak{a} \subseteq \mathcal{O}_{X,O}$ decomposes as $\mathfrak{a} = g \cdot \mathfrak{a}'$, $g \in \mathcal{O}_{X,O}$ with $\mathfrak{a}'$ being $\mathfrak{m}$-primary. The affine part F_{aff} coincides with the strict transform of the log-resolution of g so we can assume that $\mathfrak{a}$ is $\mathfrak{m}$-primary. Starting with the log-resolution of the principal ideal $(a_1 \cdots a_r)$ we blow-up some extra points: first, a finite number of free infinitely near points and, secondly, finitely many satellite points. The conditions that decide to blow-up these points are completely determined by the valuation from the log-resolution of each (a_i). Finally, the resolution becomes minimal by blowing-down unnecessary exceptional divisors, i.e., (-1)-curves. For a more detailed description, see [1, Alg. 3.14].

2.2 Computing the Integral Closure of an Ideal

Given an effective divisor $D = \sum d_i E_i \in \mathrm{Div}_\mathbb{Q}(X')$, we may consider its associated sheaf ideal $\pi_* \mathcal{O}_{X'}(-D)$. Its stalk at O is

$$H_D = \{ f \in \mathcal{O}_{X,O} \mid v_i(f) \geqslant \lceil d_i \rceil \text{ for all } E_i \leqslant D \} \tag{1}$$

where $v_i(f)$ are the coefficients of the prime divisor E_i in $\mathrm{Div}(\pi^* f)$. These ideals are *complete*, see [7], and $\mathfrak{m}$-primary whenever D has exceptional support.

Recall that an effective divisor with integral coefficients $D \in \mathrm{Div}(X')$ is called *antinef* if $-D \cdot E_i \geqslant 0$, for every exceptional prime divisor E_i. It is worth to point out that the affine part of $D = D_{\mathrm{exc}} + D_{\mathrm{aff}}$ satisfies $D_{\mathrm{aff}} \cdot E_i \geqslant 0$. Therefore D is antinef whenever $-D_{\mathrm{exc}} \cdot E_i \geqslant D_{\mathrm{aff}} \cdot E_i$. Given a non-antinef divisor D, one can compute an antinef divisor defining the same ideal, called the antinef closure, via the so called unloading procedure; see [3, Sect. 2.2] or [4, Sect. 4.6].

Algorithm 1 (Unloading procedure)
Repeat

- *define* $\Theta := \{ E_i \leqslant E \mid \rho_i = -\lceil D \rceil \cdot E_i < 0 \}$;
- *let* $n_i = \left\lceil \dfrac{\rho_i}{E_i^2} \right\rceil$ *for* $i \in \Theta$ *and notice that* $(\lceil D \rceil + n_i E_i) \cdot E_i \leqslant 0$;
- *define a new divisor as* $D' = \lceil D \rceil + \sum_{E_i \in \Theta} n_i E_i$.

Until *the resulting divisor* D' *is antinef.*

The finiteness and correctness of the unloading procedure is a consequence of the results in [3]. Antinef divisors are important when working with complete ideal since:

Theorem 2 (Lipman) *The correspondence* (1) *is one to one between antinef divisors in* $\mathrm{Div}(X')$ *and complete ideals in* $\mathcal{O}_{X,O}$, *whose log-resolution is dominated by* π.

The other essential result in the theory of complete ideals is the following:

Theorem 3 (Zariski) *Every* $\mathfrak{m}$-*primary complete ideal of* $\mathcal{O}_{X,O}$ *factors uniquely, up to order, as a product of simple complete ideals.*

A corollary of Theorems 2 and 3 is that any antinef divisor D can be decomposed uniquely, up to order, as a sum of antinef divisors D_i,

$$D = \sum \rho_i D_i, \tag{2}$$

where $\rho_i = -D \cdot E_i \geqslant 0$ and H_{D_i} is a simple ideal appearing in the Zariski factorization of H_D with multiplicity ρ_i. We call (2) the decomposition of D into simple divisors D_i.

With these results in mind, in [2] Alberich-Carramiñana–Àlvarez-Montaner–Blanco describe an algorithm to compute a system of generators for the complete ideal associated to a divisor. The algorithm is divided in two parts.

First, start with a divisor D, which we assume to be antinef. The divisor D is decomposed into simple divisors $D_1, \ldots, D_r$. For each simple divisor D_i, compute D_i' an antinef divisor defining an adjacent ideal $H_{D_i'} \subset H_{D_i}$, i.e., having codimension 1, such that D_i' is the antinef closure of $D_i + E_O$. Next, find an element $f \in \mathcal{O}_{X,o}$ belonging to H_{D_i} but not to $H_{D_i'}$. Now, D_i' is no longer simple but has smaller support than D_i. This part is repeated with $D := D_i'$ until $H_D = \mathfrak{m}$.

This first part generates a tree where each vertex is an antinef divisor and where the leafs of the tree are all $\mathfrak{m}$. The second part traverses the tree bottom-up computing in each node the ideal associated to the divisor. Using the notations from the above paragraph, given any node in the tree with divisor D, the ideal H_D is computed multiplying the ideals in child nodes $H_{D_1'} \cdots H_{D_r'}$ and adding the element f to the resulting generators. For a more detailed description, see [2, Alg. 1].

2.3 Multiplier Ideals

Let $\pi \colon X' \to X$ be a log-resolution of an ideal $\mathfrak{a} \subseteq \mathcal{O}_X$, F be the antinef divisor such that $\mathfrak{a} \cdot \mathcal{O}_{X'} = \mathcal{O}_{X'}(-F)$ and let K_π the relative canonical divisor of the resolution, in this case, it is the divisor defined by the Jacobian of π, i.e., $K_\pi = div(Jac(\pi))$. The *multiplier ideal* associated to $\mathfrak{a}$ and some real number $\lambda \in \mathbb{R}_{>0}$ is defined as

$$\mathcal{J}\left(\mathfrak{a}^\lambda\right) = \pi_* \mathcal{O}_{X'}\left(\lceil K_\pi - \lambda F \rceil\right).$$

These ideals are complete and do not depend on the log-resolution of $(X, \mathfrak{a})$. For a detailed overview of the theory of multiplier ideals and the properties they satisfy, we must refer to [3] or the book Lazarsfeld [6].

Multiplier ideals come with an attached set of invariants, which were studied systematically by Ein–Lazarsfeld–Smith–Varolin [5]. Clearly,

$$\lceil K_\pi - \lambda F \rceil \geqslant \lceil K_\pi - (\lambda + \varepsilon) F \rceil$$

for any $\varepsilon > 0$, with equality if ε is small enough. Therefore, the multiplier ideals form a discrete nested sequence of ideals

$$\mathcal{O}_{X,o} \supseteq \mathcal{J}(\mathfrak{a}^{\lambda_0}) \supsetneq \mathcal{J}(\mathfrak{a}^{\lambda_1}) \supsetneq \mathcal{J}(\mathfrak{a}^{\lambda_2}) \supsetneq \cdots \supsetneq \mathcal{J}(\mathfrak{a}^{\lambda_i}) \supsetneq \cdots$$

indexed by an increasing sequence of rational numbers $0 = \lambda_0 < \lambda_1 < \lambda_2 < \cdots$ such that, for any $c \in [\lambda_i, \lambda_{i+1})$, it holds that $\mathcal{J}(\mathfrak{a}^{\lambda_i}) = \mathcal{J}(\mathfrak{a}^c) \supsetneq \mathcal{J}(\mathfrak{a}^{\lambda_{i+1}})$. The λ_i are the so-called *jumping numbers* of the ideal $\mathfrak{a}$ and the first jumping number $\lambda_1 = \mathrm{lct}(\mathfrak{a})$ is the *log-canonical threshold* of $\mathfrak{a}$.

In Alberich-Carramiñana–Àlvarez-Montaner–Dachs-Cadefau present the following theorem that computes the consecutive jumping number to a given number by means of the unloading presented in Sect. 2.2.

Theorem 4 ([3, Thm. 3.5.]) *Let* $\mathfrak{a} \subseteq \mathcal{O}_{X,O}$ *be an ideal and let* $D_{\lambda'} = \sum e_i^{\lambda'} E_i$ *be the antinef closure of* $\lfloor \lambda' F - K_\pi \rfloor$ *for a given* $\lambda' \in \mathbb{Q}_{>0}$. *Then,*

$$\lambda = \min_i \left\{ \frac{k_i + 1 + e_i^{\lambda'}}{e_i} \right\}$$

is the jumping number consecutive to λ'.

3 Main Result

As a consequence of the previous theorem and the results already stated, one can present the following algorithm that effectively computes the chain of jumping numbers and multiplier ideals in any range.

Algorithm 5 (Jumping Numbers and Multiplier Ideals)

Input: *An ideal* $\mathfrak{a} \subseteq \mathcal{O}_{X,O}$.

Output: *The jumping numbers of* $\mathfrak{a}$ *and a set of generators for each multiplier ideal.*

(Step 1) Compute the log-resolution of $\mathfrak{a}$ *using the algorithm from Sect. 2.1.*

(Step 2) Set $\lambda_0 = 0$. *From* $j = 1$, *incrementing by 1*

(Step 2.j) · **Jumping number**: *Compute* $\lambda_j = \min_i \left\{ \frac{k_i + 1 + e_i^{\lambda_{j-1}}}{e_i} \right\}$.

· **Multiplier ideal**: *Compute the antinef closure* $D_{\lambda_j} = \sum e_i^{\lambda_j} E_i$ *of the divisor* $\lfloor \lambda_j F - K_\pi \rfloor$ *using the unloading procedure.*

· **Generators**: *Give a system of generators of the complete ideal associated to* D_{λ_j} *using the algorithm presented in Sect. 2.2.*

4   Example

Consider the ideal $\mathfrak{a} = ((y^2 - x^3)^3, x^3(y^2 - x^3)^2, x^6y^3) \subseteq \mathcal{O}_{X,O}$. We start computing the log-resolution of $\mathfrak{a}$, and we encode the information given by this process by means of several dual graphs. We represent the relative canonical divisor K_π and the divisor F in the dual graph as follows:

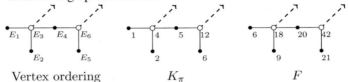

Vertex ordering K_π F

The blank dots correspond to divisors with $\rho_i > 0$. We represent this quantity by broken arrows associated to the divisors (these broken arrows also represent the branches of the strict transform of a curve defined by a generic $f \in \mathfrak{a}$). For simplicity we will collect the values of any divisor in a vector. To begin with, we have $K_\pi = (1, 2, 4, 5, 6, 12)$ and $F = (6, 9, 18, 20, 21, 42)$. We can now perform the second step of the algorithm. For the sake of space, we have selected three multiplier ideals:

- The first jumping number, the log canonical threshold, is $5/18$, and the multiplier ideal associated is $\mathcal{J}\left(\mathfrak{a}^{5/18}\right) = (x, y)$.
- The first jumping number not having a monomial ideal as a multiplier ideal is the one associated to $\lambda_6 = 25/42$. The divisor $\lfloor \frac{25}{42}F - K_\pi \rfloor = (2, 3, 6, 6, 6, 13)$ is not antinef. So we need to perform several unloading steps to obtain the antinef closure $D_{\lambda_6} = (2, 3, 6, 7, 7, 14)$. The associated multiplier ideal is $\mathcal{J}\left(\mathfrak{a}^{25/42}\right) = (y^2 - x^3, x^2y, xy^2, x^4)$.
- The biggest jumping number smallest than one, the one defining the so called adjoint ideal (see [6]), is $\lambda_{19} = 41/42$. The divisor $\lfloor \frac{41}{42}F - K_\pi \rfloor = (4, 6, 13, 14, 14, 29)$ is not antinef. Hence, we need to perform an unloading step to obtain the antinef closure $D_{\lambda_{19}} = (5, 7, 13, 14, 15, 29)$. Finally, the associated multiplier ideal is

$$\mathcal{J}\left(\mathfrak{a}^{\frac{41}{42}}\right) = \left(x^8, x^6y, x^5y^2, x^4(y^2 - x^3),\right.$$
$$\left. x^3y^3, x^2y^4, x^2y(y^2 - x^3), xy^2(y^2 - x^3), y^3(y^2 - x^3)\right).$$

References

1. M. Alberich-Carramiñana, J. Àlvarez-Montaner, G. Blanco, *Effective Computation of Base Points of Ideals in Two-Dimensional Local Rings*, arXiv:1605.05665
2. M. Alberich-Carramiñana, J. Àlvarez-Montaner, G. Blanco, *Monomial generators of complete planar ideals*, arXiv:1701.03503v3

3. M. Alberich-Carramiñana, J. Àlvarez-Montaner, F. Dachs-Cadefau, Multiplier ideals in two-dimensional local rings with rational singularities. Mich. Math. J. **65**(2), 287–320 (2016)
4. E. Casas-Alvero, *Singularities of Plane Curves*. London Mathematical Society Lecture Note Series, vol. 276 (Cambridge University Press, Cambridge, 2000)
5. L. Ein, R. Lazarsfeld, K. Smith, D. Varolin, Jumping coefficients of multiplier ideals. Duke Math. J. **123**, 469–506 (2004)
6. R. Lazarsfeld, *Positivity in Algebraic Geometry. II*, vol. 49, (Springer, Berlin, 2004) pp. xviii+385
7. O. Zariski, Polynomial ideals defined by infinitely near points. Am. J. Math. **60**(1), 151204 (1938)

Notes on Divisors Computing MLD's and LCT's

Harold Blum

Abstract This note discusses results presented at the 2016 meeting "Workshop on Positivity and Valuations" at Centre de Recerca Matemàtica. Much of the content discussed below appears in Blum (On divisors computing mld's and lct's, 2016, [1]) with further details.

1 Introduction

The *log canonical threshold* and *minimal log discrepancy* are two invariants of singularities that arise naturally in the study of birational geometry. Minimal log discrepancies are of particular interest due to a work of Shokoruv [11] in which he proved that two conjectures on minimal log discrepancies (semicontinuity and the ascending chain condition (ACC)) imply the termination of flips, a result needed to complete the minimal model program in full generality.

Shokurov originally conjectured that both the set of minimal log discrepancies and log canonical thresholds in fixed dimension should satisfy the ACC. The conjecture was proved for log canonical thresholds on smooth varieties in [3] and later in full generality [5]. The general form of the ACC conjecture for minimal log discrepancies remains open. In this way, as well as others, minimal log discrepancies are less well understood than log canonical thresholds.

In order to define these two invariants, we will recall the following notions. Fix a variety X over an algebraically closed field of characteristic 0. A *divisor over* a variety X corresponds to a prime divisor on a normal variety Y, proper and birational over X. That is to say, E is a divisor over X if there exists a projective birational morphism $f : Y \to X$ with Y normal and $E \subseteq Y$ a prime divisor. Since Y is normal,

I would like to thank my advisor Mircea Mustaţă for introducing me to many of the topics discussed in this note and guiding my research that resulted in [1]. I would also like to thank Mattias Jonsson and Karen Smith for interesting and useful conversations.

H. Blum (✉)
Department of Mathematics, University of Michigan, Ann Arbor, MI 48109, USA
e-mail: blum@umich.edu

© Springer Nature Switzerland AG 2018
M. Alberich-Carramiñana et al. (eds.), *Extended Abstracts February 2016*,
Trends in Mathematics 9, https://doi.org/10.1007/978-3-030-00027-1_10

$\mathcal{O}_{Y,E}$ is a discrete valuation ring. We write ord_E for the corresponding valuation of $K(X)$.

We call $(X, \mathfrak{a}^\lambda)$ a *pair* if X is a normal **Q**-Gorenstein variety, $\mathfrak{a} \subseteq \mathcal{O}_X$ a nonzero ideal, and $\lambda \in \mathbf{R}_{\geq 0}$. The *log discrepancy* of a pair $(X, \mathfrak{a}^\lambda)$ along E is defined as

$$a_E\left(X, \mathfrak{a}^\lambda\right) := k_E + 1 - \lambda \operatorname{ord}_E(\mathfrak{a}),$$

where k_E is the coefficient of E in the relative canonical divisor $K_{Y/X}$ and ord_E is the valuation as mentioned above. A pair $(X, \mathfrak{a}^\lambda)$ is klt (resp., log canonical) if for all divisors E over X, $a_E(X, \mathfrak{a}^\lambda) > 0$ (resp., ≥ 0). A normal **Q**-Gorenstein variety is said to be klt if $(X, \mathcal{O}_X)$ is a klt pair. This is equivalent to requiring X have a log resolution $\tilde{X} \to X$ such that the coefficients of $K_{\tilde{X}/X}$ are strictly larger than -1.

Arising from these definitions are two invariants that measure the "nastiness" of a singularity. Assuming X has klt singularities, the *log canonical threshold* of a nonzero ideal $\mathfrak{a}$ on X is defined as

$$\operatorname{lct}(\mathfrak{a}) := \sup\{\lambda \in \mathbf{R}_{\geq 0} | \, (X, \mathfrak{a}^\lambda) \text{ is log canonical}\}.$$

Alternatively, it is straightforward to show that

$$\operatorname{lct}(\mathfrak{a}) := \min\left\{\frac{k_E + 1}{\operatorname{ord}_E(\mathfrak{a})} \mid E \text{ is a divisor over } X\right\}.$$

Given a klt pair $(X, \mathfrak{a}^\lambda)$ and a (not necessarily closed) point $\eta \in X$, the *minimal log discrepancy* of $(X, \mathfrak{a}^\lambda)$ at η is defined as

$$\operatorname{mld}_\eta(X, \mathfrak{a}^\lambda) = \min\{a_E(X, \mathfrak{a}^\lambda) \mid E \text{ a divisor over } X \text{ with } c_X(E) = \overline{\eta}\}.$$

See [1] for further details on these definitions.

In understanding these two invariants it is natural to make the following definition. Given a divisor E over X, we say that E *computes a log canonical threshold* if there exists a nonzero ideal $\mathfrak{a}$ on X such that $a_E\left(X, \mathfrak{a}^\lambda\right) = 0$, where $\lambda = \operatorname{lct}(\mathfrak{a})$. Furthermore, we say that E *computes* $\operatorname{lct}(\mathfrak{a})$. Similarly, we say that E *computes a minimal log discrepancy* if there exists a pair $(X, \mathfrak{a}^\lambda)$ such that $\operatorname{mld}_\eta(X, \mathfrak{a}^\lambda) = a_E\left(X, \mathfrak{a}^\lambda\right)$ with $\overline{\eta} = c_X(E)$. Furthermore, we say that E *computes* $\operatorname{mld}_\eta(X, \mathfrak{a}^\lambda)$.

Question 1 *Which divisors over a variety compute log canonical thresholds (resp., minimal log discrepancies)?*

Divisors computing log canonical thresholds satisfy special properties. As we will explain shortly, it is well known that divisors computing log canonical thresholds have finitely generated graded sequences of ideals. It is not known if the same can be said for divisors computing minimal log discrepancies. Note that if E computes $\lambda = \operatorname{lct}(\mathfrak{a})$ and $\overline{\eta} = c_X(E)$, then E also computes $\operatorname{mld}_\eta(X, \mathfrak{a}^\lambda)$, which is 0. Thus, divisors computing log canonical thresholds also compute minimal log discrepancies. The reverse statement does not necessarily hold.

2 Smooth Surfaces

When our variety is a smooth surface, Question 1 has the following simple answer.

Theorem 2 *If X is a smooth surface, then every divisor over X centered at a point that computes a minimal log discrepancy also computes a log canonical threshold.*

It is known which divisors satisfy the hypotheses of the theorem. If E is a divisor over a smooth surface X with $c_X(E) = \{x\}$ such that E computes a log canonical threshold, then ord_E is a monomial valuation in some analytic coordinates at x [4, Lem. 2.11]. For further results on divisors computing log canonical thresholds on surfaces, see [12, 13].

A key ingredient in our proof of Theorem 2 is to look at an invariant that will determine which divisors compute log canonical thresholds. We will examine this invariant in the following section.

3 Relation to the Log Canonical Threshold of a Graded Sequence

We recall that a *graded sequence of ideals* on X is a sequence of ideals $\mathfrak{a}_\bullet = \{\mathfrak{a}_m\}_{m \in \mathbf{N}}$ on X such that $\mathfrak{a}_m \cdot \mathfrak{a}_n \subseteq \mathfrak{a}_{m+n}$ for all $m, n \in \mathbf{N}$. Assuming $\mathfrak{a}_0 = \mathcal{O}_X$, we say that $\mathfrak{a}_\bullet$ is finitely generated if the Rees algebra

$$R(\mathfrak{a}_\bullet) = \bigoplus_{m \in \mathbf{N}} \mathfrak{a}_m$$

is a finitely generated $\mathcal{O}_X$-algebra. If X is a klt variety and $\mathfrak{a}_\bullet$ is a graded sequence of ideals on X such that $\mathfrak{a}_m$ is not the zero ideal for all $m \in \mathbf{N}$, then we may define

$$\mathrm{lct}(\mathfrak{a}_\bullet) := \lim_{m \to \infty} m \cdot \mathrm{lct}(\mathfrak{a}_m).$$

See [7] for further details on this invariant.

A divisor over a variety X gives rise to a graded sequence of ideals as follows. Let $f : Y \to X$ be a proper birational morphism of normal varieties and $E \subseteq Y$ be a prime divisor. We write $\mathfrak{a}_\bullet^E = \{\mathfrak{a}_m^E\}_{m \in \mathbf{N}}$, where

$$\mathfrak{a}_m^E := f_* \mathcal{O}_Y(-mE).$$

Locally, $\mathfrak{a}_m^E$ can be expressed as functions on X vanishing to at least order m along E.

Proposition 3 *If E is a divisor over a klt variety X, then E computes a log canonical threshold if and only if* $\mathrm{lct}(\mathfrak{a}_\bullet^E) = k_E + 1$.

For a divisor E over X, the invariant $\mathrm{lct}(\mathfrak{a}_\bullet^E)$ has a natural geometric interpretation related to the finite generation of $\mathfrak{a}_\bullet^E$. First, we note the following interpretation of the finite generation of $\mathfrak{a}_\bullet^E$.

Theorem 4 ([6, Cor. 3.3]) *Let X be a normal variety and E a divisor over X such that $\mathrm{codim}(c_X(E)) \geq 2$. If $\mathfrak{a}_\bullet^E$ is finitely generated, then $\mathrm{Proj}_X(\oplus_{m \geq 0} \mathfrak{a}_m^E) \to X$ is a proper birational morphism of normal varieties with exactly one exceptional divisor and the divisor corresponds to E.*

Such birational morphisms with exactly one exceptional divisor were studied in [6] and referred to as *prime blowups*. In [10], the author looked at *plt-blowups* which are prime blowups with restrictions on the singularities of the blowup. The following proposition relates the value of $\mathrm{lct}(\mathfrak{a}_\bullet^E)$ to the model $\mathrm{Proj}_X(\oplus_{m \geq 0} \mathfrak{a}_m^E)$ in Theorem 4.

Proposition 5 *If X is a klt variety and E a divisor over X with $k_E < \mathrm{lct}(\mathfrak{a}_\bullet^E)$ then,*

 (i) *the graded sequence $\mathfrak{a}_\bullet^E$ is finitely generated, and*
 (ii) *the model $Y := \mathrm{Proj}_X(\oplus_{m \geq 0} \mathfrak{a}_m^E)$ (see Theorem 4) has klt singularities with $\mathrm{lct}(Y, E_Y) = \mathrm{lct}(\mathfrak{a}_\bullet^E) - k_E$, where E_Y is the prime divisor on Y identified with E.*

The first assertion of the above proposition is an elegant restatement of the fact that if a divisor E has log discrepancy in the interval $[0,1)$ along a klt pair, then $\mathfrak{a}_\bullet^E$ is finitely generated. (When X is $\mathbf{Q}$-factorial, this follows directly from [8, Cor. 1.39].)

In the literature, there are two examples of divisors over smooth varieties with non-finitely generated graded sequences of ideals [2, 9]. Again, by [8, Cor. 1.39], it is well known that these divisors cannot compute log canonical thresholds (but not whether they compute minimal log discrepancies). By furthering understanding information related to $\mathrm{lct}(\mathfrak{a}_\bullet^E)$, [1] explains that neither divisor computes a minimal log discrepancy.

References

1. H. Blum, On divisors computing mld's and lct's. (2016), arXiv:1605.09662
2. V. Cossart, C. Galindo, O. Piltant, Un exemple effectif de gradué non Noethérien associé à une valuation divisorielle. Ann. Inst. Fourier (Grenoble) **50**(1), 105–112 (2000)
3. T. de Fernex, L. Ein, M. Mustaţă, Shokurov's ACC conjecture for log canonical thresholds on smooth varieties. Duke Math. J. **152**(1), 93–114 (2010)
4. C. Favre, M. Jonsson, Valuations and multiplier ideals. J. Am. Math. Soc. **18**(3), 655–684 (2005) (electronic)
5. C.D. Hacon, J. McKernan, C. Xu, ACC for log canonical thresholds. Ann. Math. **180**(2), 523–571 (2014)
6. S. Ishii, Extremal functions and prime blow-ups. Commun. Algebra **32**(3), 819–827 (2004)
7. M. Jonsson, M. Mustaţă, Valuations and asymptotic invariants for sequences of ideals. Ann. Inst. Fourier (Grenoble) **62**(6), 2145–2209 (2013)

8. J. Kollár, *Singularities of the Minimal Model Program*, vol. 200, Cambridge Tracts in Mathematics (Cambridge University Press, Cambridge, 2013)
9. A. Küronya, A divisorial valuation with irrational volume. J. Algebra **262**(2), 413–423 (2003)
10. Y.G. Prokhorov, Blow-ups of canonical singularities. in *Algebra (Moscow, 1998)* (de Gruyter, Berlin, 2000), pp. 301–317
11. V.V. Shokurov, Letters of a bi-rationalist. V. Minimal log discrepancies and termination of log flips. Tr. Mat. Inst. Steklova **246**, 328–351, (2004) (Algebr. Geom. Metody, Svyazi i Prilozh.)
12. K.E. Smith, H.M. Thompson, Irrelevant exceptional divisors for curves on a smooth surface, in *Algebra, Geometry and their Interactions*, vol. 448 Contemporary Mathematics (American Mathematical Society, Providence, 2007), pp 245–254
13. K. Tucker, Jumping numbers on algebraic surfaces with rational singularities. Trans. Am. Math. Soc. **362**(6), 3223–3241 (2010)

On the Containment Hierarchy for Simplicial Ideals

Magdalena Lampa-Baczyńska and Grzegorz Malara

Abstract The containment relations between symbolic and ordinary powers of homogeneous ideals recently became a popular direction of research in algebraic geometry. Our considerations are mainly inspired by results of Bocci–Harbourne in (Bocci and Harbourne. Proc. Am. Math. Soc. 138, 1175–1190 (2010) [1]). The results presented on a poster are from (Lampa-Baczyńska and Malara. J. Pure Appl. Algeb. 219, 5402–5412 (2015) [2]), the common paper of both authors.

1 Preliminaries

We begin by introducing a new object, the *simplicial ideal*, which is the main subject of our considerations.

Definition 1 (*Simplicial ideal*) A *simplicial ideal* is the ideal of the codimension c skeleton of the simplex spanned by all coordinate points in $\mathbb{P}^n$ denoted by $I(n, c)$.

More exactly, if H_i is the hyperplane $\{x_i = 0\}$ for $i = 0, \ldots, n$, then the set of zeroes of $I(n, c)$ is the union of all c-fold intersections $H_{i_1} \cap \cdots \cap H_{i_c}$ for mutually distinct indices $i_1, \ldots, i_c \in \{0, \ldots, n\}$.

Let us now recall some general mathematical objects and its properties. We assume that we work over an arbitrary field $\mathbb{K}$. Denote by $S(n) = \mathbb{K}[x_0, \ldots, x_n]$ the ring of polynomials over $\mathbb{K}$.

Definition 2 Let $I \subseteq S(n)$ be a homogeneous ideal and let $m \geq 1$ be a positive integer. The *m-th symbolic power* of I is

$$I^{(m)} = S(n) \cap \Big(\bigcap_{Q \in \mathrm{Ass}(I)} I_Q^m \Big),$$

M. Lampa-Baczyńska (✉) · G. Malara
Pedagogical University of Kraków, Kraków, Poland
e-mail: lampa.baczynska@wp.pl

G. Malara
e-mail: gmalara@up.krakow.pl

© Springer Nature Switzerland AG 2018
M. Alberich-Carramiñana et al. (eds.), *Extended Abstracts February 2016*,
Trends in Mathematics 9, https://doi.org/10.1007/978-3-030-00027-1_11

where $\mathrm{Ass}(I)$ is the set of all associated primes Q of I, and I_Q denotes the localization of I in Q. The intersection takes place in the field of fractions of $S(n)$.

Although symbolic powers are defined algebraically, they have a nice geometrical interpretation. Let us recall the Nagata–Zariski Theorem (see [3, Thm. 3.14], and [4, Cor. 2.9]) which puts the previous definition into another perspective.

Theorem 3 (Nagata–Zariski) *Let $I \subseteq S(n)$ be a radical ideal and let V be the set of zeroes of I. Then, $I^{(m)}$ consists of all polynomials vanishing to order at least m along V.*

2 The Containment Problem

The containment problem for symbolic and usual powers of ideals has been intensively studied in recent years; see, e.g., [1, 5, 6]. In all these works the authors study containment relations of the type $I^{(m)} \subset I^r$ for *a fixed* homogeneous ideal I. There appear many conjectures about this type and other similar types of containments; see [7] for details.

For an arbitrary homogeneous ideal I Bocci and Harbourne introduced in [5] a quantity $\rho(I)$ called resurgence, which is an asymptotic invariant of I, very helpful in the research area of containment relations.

Definition 4 Let $I \subseteq S(n)$ be a non-trivial (i.e., $I \neq \langle 0 \rangle$ and $I \neq \langle 1 \rangle$) homogeneous ideal. *The resurgence* of I is the real number

$$\rho(I) := \sup\left\{\frac{m}{r} : I^{(m)} \nsubseteq I^r\right\}.$$

This invariant is of interest as it guarantees the containment $I^{(m)} \subseteq I^r$, for $m/r > \rho(I)$.

There are obvious inclusions which follow from the definitions and the Nagata–Zariski Theorem, namely, $I = I^{(1)} \supseteq I^{(2)} \supseteq I^{(3)} \supseteq \cdots$ and $I = I^1 \supseteq I^2 \supseteq I^3 \supseteq \cdots$.

It is natural to ask for what values of m and r we have the inclusions (i) $I^r \subseteq I^{(m)}$ and (ii) $I^{(m)} \subseteq I^r$. It is easy to see that (i) holds if and only if $m \leq r$.

As for (ii), Ein–Lazarsfeld–Smith [8] in characteristic zero, and Hochster–Huneke [9] in arbitrary characteristic, showed that there is always containment for $m \geq nr$. Of course, in certain cases (e.g., complete intersections ideals) this bound is not optimal and the problem has to be studied individually in any given case. It might be worth to mention that in fact it is not known if the bound $m \geq nr$ is ever optimal, i.e., no ideal I is known such that the containment $I^{(m)} \subseteq I^r$ requires $m \geq nr$ for all r.

3 Main Results

Here we present the following relations between symbolic and ordinary powers of simplicial ideals. These results come from [2]. There one can also see detailed proofs of theorems. Here we present only their statements.

Theorem 5 *For $n \geq 1$ and $c \in \{1, \ldots, n\}$, there is the containment $I^{(m)}(n, c) \subset I^r(n, c)$ if and only if*

$$r \leq \frac{(n+1)k - p}{n - c + 2},$$

where $m = kc - p$ and $0 \leq p < c$.

In fact our approach is a little bit more general. It is motivated by the obvious containments hierarchy

$$I(n, 1) \subset I(n, 2) \subset \cdots \subset I(n, n-1) \subset I(n, n).$$

Thus we extend the containment problem to the inclusion relations between symbolic powers of *various* simplicial ideals. Our main result in this direction is the following.

Theorem 6 *Let n be a positive integer and let $c, d \in \{1, \ldots, n\}$. If $c \leq d$ and $sc \leq md$ then there is the containment $I^{(m)}(n, c) \subset I^{(s)}(n, d)$.*

The resurgence $\rho(I)$ measures in effect the asymptotic discrepancy between symbolic and ordinary powers of a given ideal. This is a delicate invariant and the family of ideals for which its value is known is growing rather slowly; see, e.g., [10]. Here, we expand this knowledge a little bit.

Theorem 7 *For a positive integer n and $c \in \{1, \ldots, n\}$ the following holds*

$$\rho(I(n, c)) = \frac{c(n + 2 - c)}{n + 1}.$$

Notice that the value for the resurgence claimed in the expression above was known to be an upper bound [5, Thm. 2.4.3 b] and that the case $I(n, n)$ was already computed in [5, Thm. 2.4.3 a].

References

1. C. Bocci, B. Harbourne, The resurgence of ideals of points and the containment problem. Proc. Am. Math. Soc. **138**, 1175–1190 (2010)
2. M. Lampa-Baczyńska, G. Malara, On the containment hierarchy for simplicial ideals. J. Pure Appl. Algeb. **219**, 5402–5412 (2015)
3. D. Eisenbud, *Commutative Algebra. With a View Toward Algebraic Geometry*, (Springer, New York, 1995)

4. J. Sidman, S. Sullivant, Prolongations and computational algebra. Can. J. Math. **61**, 930–949 (2009)
5. C. Bocci, B. Harbourne, Comparing powers and symbolic powers of ideals. J. Algebr. Geom. **19**, 399–417 (2010)
6. B. Harbourne, C. Huneke, Are symbolic powers highly evolved? J. Ramanujan Math. Soc. **28**, 311–330 (2013)
7. M. Dumnicki, T. Szemberg, H. Tutaj-Gasińska, Counterexamples to the $I^{(3)} \subset I^2$ containment. J. Algeb. **393**, 24–29 (2013)
8. L. Ein, R. Lazarsfeld, K.E. Smith, Uniform bounds and symbolic powers on smooth varieties. Invent. Math. **144**, 241–252 (2001)
9. M. Hochster, C. Huneke, Comparison of symbolic and ordinary powers of ideals. Invent. Math. **147**, 349–369 (2002)
10. M. Dumnicki, B. Harbourne, U. Nagel, A. Seceleanu, T. Szemberg, H. Tutaj-Gasińska, Resurgences for ideals of special point configurations in $\mathbb{P}^N$ coming from hyperplane arrangements. J. Algeb. **443**, 383–394 (2015)

The Universal Zeta Function for Curve Singularities and its Relation with Global Zeta Functions

Julio José Moyano-Fernández

Abstract The purpose of this note is to give a brief overview on zeta functions of curve singularities and to provide some evidences on how these and global zeta functions associated to singular algebraic curves over perfect fields relate to each other.

1 Introduction

Let X be a complete, geometrically irreducible, singular algebraic curve defined over a perfect field k; from now on we will refer to such a curve simply as 'algebraic curve over k'. Let K be the field of rational functions on X. Extending previous works of V. M. Galkin and B. Green—and based on the classical results of Schmidt [1] for nonsingular curves—K.O. Stöhr (cf. [2, 3]) managed to attach a zeta function to X for finite k in the following manner: If $\mathcal{O}_X$ is the structure sheaf of X, he defined the Dirichlet series

$$\zeta(\mathcal{O}_X, s) := \sum_{\mathfrak{a} \geq \mathcal{O}_X} q^{-s \deg \mathfrak{a}}, \quad s \in \mathbb{C} \text{ with } \operatorname{Re}(s) > 0,$$

where the sum is taken over all positive divisors of X, and $\deg(\,\cdot\,)$ denotes the degree of those divisors. Observe that the change of variables $T = q^{-s}$ allows to consider the formal power series in T

The author was partially supported by the Spanish Government Ministerio de Economía, Industria y Competitividad (MINECO), grants MTM2012-36917-C03-03 and MTM2015-65764-C3-2-P, as well as by Universitat Jaume I, grant P1-1B2015-02.

J. J. Moyano-Fernández (✉)
Universitat Jaume I, Campus de Riu Sec, Departament de Matemàtiques
& Institut Universitari de Matemàtiques i Aplicacions de Castelló,
12071 Castellón de la, Plana, Spain
e-mail: moyano@uji.es

© Springer Nature Switzerland AG 2018
M. Alberich-Carramiñana et al. (eds.), *Extended Abstracts February 2016*,
Trends in Mathematics 9, https://doi.org/10.1007/978-3-030-00027-1_12

$$Z(\mathcal{O}_X, T) = \sum_{n=0}^{\infty} \#(\{\text{positive divisors of } X \text{ of degree } n\}) \cdot T^n.$$

Moreover, Stöhr considered local zeta functions, i.e., zeta functions attached to every local ring $\mathcal{O}_P$ of points P at X of the form

$$Z(\mathcal{O}_P, T) := \sum_{\mathfrak{a} \supseteq \mathcal{O}_P} T^{\deg \mathfrak{a}} = \sum_{n=0}^{\infty} \#(\{\text{positive } \mathcal{O}_P-\text{ideals of degree } n\}) \cdot T^n.$$

This series extends previous definitions by Galkin [4] and Green [5]. Furthermore, the Euler product formula for the formal power series yields the identity

$$Z(\mathcal{O}_X, T) = \prod_{P \in X} Z(\mathcal{O}_P, T),$$

which actually establishes a link between the local and global theory. Every local factor $Z(\mathcal{O}_P, T)$ splits again into factors

$$Z(\mathcal{O}_P, \mathcal{O}_P, T) = \sum_{n=0}^{\infty} \#(\{\text{principal integral } \mathcal{O}-\text{ideals of codimension } n\}) \cdot T^n$$

which are determined by the value semigroup of $\mathcal{O}_P$ (see Sect. 2 below for the definition of this semigroup) if the field is big enough, as Zúñiga showed in [6].

On the other hand, when studying the Gorenstein property of one-dimensional local Cohen–Macaulay rings, Campillo–Delgado–Kiyek [7, Sect. 3.8] observed the existence of a Laurent series—a polynomial in their situation—attached to those rings, and satisfying a functional equation in the case of Gorenstein rings. Further investigations by Campillo, Delgado and Gusein-Zade [8–14] led to the definition of a Poincaré series associated to a complex curve singularity as an integral with respect to the Euler characteristic; see also Viro [15]. They even considered integration with respect to an Euler characteristic of motivic nature and so they introduced the notion of *generalized Poincaré series* of a complex curve singularity [16].

In the spirit of the preceding paragraphs, the author showed in his thesis [17] (see also the joint paper with his advisor Delgado [18]) that the factors $Z(\mathcal{O}_P, \mathcal{O}_P, T)$ coincide essentially with the generalized Poincaré series of Campillo, Delgado and Gusein-Zade, under a suitable specialization for finite fields (see Remark 6 below). These ideas have also provided some feedback: for instance Stöhr achieved a deeper insight into the nature of the local zeta functions (see [19, 20] together with his student J.J. Mira).

The key ingredient that allows to relate those different formal power series is the *universal zeta function* for a curve singularity defined by Zúñiga and the author in [21]: for example, the local zeta functions and Poincaré series mentioned above are specializations of this universal zeta function. After some preliminaries, we devote Sect. 3 to describe this series. Moreover, we claim that one may establish the local-global behaviour explained above for curves defined over non-finite fields. This conjectural behaviour has already shown some evidences in particular cases; see e.g. the theorem in Sect. 4.

2 Preliminaries and Notation

Consider the normalization $\pi\colon \tilde{X} \to X$ of an algebraic curve X over k, and let $\mathcal{O} = \mathcal{O}_P := \mathcal{O}_{P,X}$ be the local ring of X at P. For the sake of simplicity we will assume the ring $\mathcal{O}$ to be complete.

It is $\pi^{-1}(P) = \{Q_1, \ldots, Q_d\}$ and so the corresponding local rings $\mathcal{O}_{Q_i}$ are discrete valuation rings of K over $\mathcal{O}$. The value semigroup associated to $\mathcal{O}$ is defined to be

$$S(\mathcal{O}) := \{\underline{v}(\underline{z}) : \underline{z} \text{ nonzero divisor in } \mathcal{O}\} \subseteq \mathbb{N}^d;$$

here $\underline{v}(\underline{z}) = (v_1(z_1), \ldots, v_d(z_d))$, where each v_i stands for the valuation associated with $\mathcal{O}_{Q_i}$; we write S for this semigroup from now on. Let $c = c(S)$ denote the conductor of S, i.e., the smallest element $\underline{v} \in S$ such that $\underline{v} + \mathbb{N}^d \subseteq S$. Moreover, $\mathcal{O}^\times$ denotes the group of units of $\mathcal{O}$. Further details here and in the sequel can be checked in [21] and the references therein.

We say that the ring $\mathcal{O}$ is totally rational if all rings $\mathcal{O}_{Q_i}$, for $i = 1, \ldots, d$ have k as a residue field.

The integral closure of $\mathcal{O}$ in K/k is $\tilde{\mathcal{O}} = \tilde{\mathcal{O}}_P = \mathcal{O}_{Q_1} \cap \cdots \cap \mathcal{O}_{Q_d}$. We write $\tilde{\mathcal{O}}^\times$ for its group of units. The *singularity degree* δ_P of $\tilde{\mathcal{O}}$ is defined as $\delta_P = \delta := \dim_k \tilde{\mathcal{O}}/\mathcal{O} < \infty$; see e.g. [22, Ch. IV].

For $\underline{n} \in S$ we set $\mathcal{I}_{\underline{n}} := \{I \subseteq \mathcal{O} \mid I = \underline{z}\mathcal{O}, \text{ with } \underline{v}(\underline{z}) = \underline{n}\}$ and, for $m \in \mathbb{N}$,

$$\mathcal{I}_m := \bigcup_{\substack{\underline{n} \in S \\ \|\underline{n}\| = m}} \mathcal{I}_{\underline{n}},$$

where $\|\underline{n}\|$ denotes the sum of the components of the vector $\underline{n} = (n_1, \ldots, n_d) \in \mathbb{N}^d$.

We define the Grothendieck ring $K_0(\mathrm{Var}_k)$ in the category Var_k of k-algebraic varieties, as the ring generated by symbols $[V]$ for $V \in \mathrm{Var}_k$, with the relations $[V] = [W]$ if V is isomorphic to W, $[V] = [V \setminus Z] + [Z]$ if Z is closed in V, and $[V \times W] = [V][W]$. We write $\mathbb{L} := [\mathbb{A}_k^1]$ for the class of the affine line, and $\mathcal{M}_k := K_0(\mathrm{Var}_k)[\mathbb{L}^{-1}]$ for the ring obtained by localization with respect to the multiplicative set generated by $\mathbb{L}$.

It is possible to associate to $\mathcal{I}_{\underline{n}}$ resp. $\mathcal{I}_m$ well-defined classes in the Grothendieck ring [21, Sect. 5]; those classes will be denoted by $[\mathcal{I}_{\underline{n}}]$ resp. $[\mathcal{I}_m]$. This allows to attach to the local ring $\mathcal{O}$ the zeta functions

$$Z(T_1, \ldots, T_d, \mathcal{O}) := \sum_{\underline{n} \in S} [\mathcal{I}_{\underline{n}}] \mathbb{L}^{-\|\underline{n}\|} T^{\underline{n}} \in \mathcal{M}_k[\![T_1, \ldots, T_d]\!],$$

where $T^{\underline{n}} := T_1^{n_1} \cdot \cdots \cdot T_d^{n_d}$, and $Z(T, \mathcal{O}) := Z(T, \ldots, T, \mathcal{O})$.

Definition 1 Consider an algebraic curve X over k. If k has characteristic $p \geq 0$, then we say that k *is big enough for* X if for every singular point P in X the following two conditions hold: (i) the ring $\mathcal{O}$ is totally rational; and (ii) $\widetilde{\mathcal{O}}^{\times} / \mathcal{O}^{\times} \cong (G_m)^{d-1} \times (G_a)^{\delta - d + 1}$, with $G_m = (k^{\times}, \cdot)$ and $G_a = (k, +)$.

Note that the condition 'k is big enough for X' is fulfilled when p is big enough.

3   The Universal Zeta Function for Curve Singularities

For $k = \mathbb{C}$, we consider a semigroup $S \subseteq \mathbb{N}^d$ such that $S = S(\mathcal{O})$. Moreover, for $\underline{n} \in S$ set

$$\mathcal{I}_{\underline{n}}(U) := (U - 1)^{-1} U^{\|\underline{n}\| + 1} \sum_{I \subseteq [d]} (-1)^{\#(I)} U^{-\dim_k \left(\mathcal{O} / \{\underline{z} \in \mathcal{O} : \underline{v}(\underline{z}) \geqslant \underline{n} + \underline{1}_I\} \right)},$$

for an indeterminate U, and where $[d] := \{1, 2, \dots, d\}$, and $\underline{1}_I$ is the d-tuple with the components corresponding to the indices in I equal to 1, and the other components equal to 0.

The notation $\mathcal{I}_{\underline{n}}(U)$ is appropriate, since that expression coincides with $\left[\mathcal{I}_{\underline{n}} \right]$ when U specializes to $\mathbb{L}$, cf. [21, Sect. 5].

Let $\underline{c} = (c_1, \dots, c_d)$ be the conductor of the semigroup S, see Sect. 2. Let $J := \{1, \dots, r\} \subseteq [d]$, and let $\underline{m} \in \mathbb{N}^d$ be such that $\underline{c} > \underline{m}$, i.e., $c_i > m_i$ for all $i \in [d]$. For a fixed $\emptyset \subsetneq J \subsetneq [d]$, set $r_J := \#J$ and $B_J := \{\underline{m} \in \mathbb{N}^{r_J} \mid H_{J, \underline{m}} \neq \emptyset\}$, where $H_{J, \underline{m}} := \{\underline{n} \in S : n_j \geq c_j \text{ if } j \in J, \text{ and } n_j = m_j \text{ otherwise}\}$.

Definition 2 We define the *universal zeta function* $\mathcal{Z}(T_1, \dots, T_d, U, S)$ associated with S to be

$$\sum_{\substack{\underline{n} \in S \\ \underline{0} \leqslant \underline{n} < \underline{c}}} \mathcal{I}_{\underline{n}}(U) U^{-\|\underline{n}\|} T^{\underline{n}} + \sum_{\emptyset \subsetneq J \subsetneq I_0} \sum_{\underline{m} \in B_J} (U - 1) U^{\|\underline{c}\| - \delta - 1} \mathcal{I}_{f_J(\underline{m})}(U) U^{-\|\underline{c}\| - \left\| \underline{f_J(\underline{m})} \right\|} \times$$

$$\times \frac{T^{\underline{f_J(\underline{m})}}}{\prod_{i=1}^{r_J} \left(1 - U^{-1} T_i \right)} + \frac{(U - 1)^{d-1} U^{\delta - d + 1} U^{-\|\underline{c}\|} T^{\underline{c}}}{\prod_{i=1}^{d} \left(1 - U^{-1} T_i \right)},$$

where $\underline{f_J(\underline{m})} = \left(c_1, \dots, c_{r_J}, m_{r_J + 1}, \dots, m_d \right) \in S$, with $m_i < c_i$, $r_J + 1 \leqslant i \leq d$, and $1 \leqslant r_J < d$.

Observe that this universal zeta function is completely determined by S. The adjective *universal* applied to this zeta function will be clear after the following paragraphs.

The generalized Poincaré series $P_g(T_1, \dots, T_d)$ of Campillo, Delgado and Gusein-Zade ([16]; see also [7, 18]) as an integral with respect to an Euler characteristic of motivic nature is very close to the zeta function $Z(T_1, \ldots, T_d, \mathcal{O})$ from Sect. 2, and therefore to the universal zeta function via the specialization $U = \mathbb{L}$:

Proposition 3 *If $S = S(\mathcal{O})$ and k is big enough for Y then,*

$$Z(T_1, \ldots, T_d, \mathcal{O}) = \mathbb{L}^{\delta+1} P_g(T_1, \ldots, T_d) = \mathcal{Z}(T_1, \ldots, T_d, U, S)|_{U=\mathbb{L}}.$$

In addition, a certain specialization of the universal zeta function coincides with the zeta function of the monodromy transformation of a reduced plane curve singularity acting on its Milnor fibre, as we briefly explain now.

Definition 4 Let $(X, 0) \subseteq (\mathbb{C}^2, 0)$ be a reduced plane curve singularity defined by an equation $f = 0$, with $f \in \mathcal{O}_{(\mathbb{C}^2, 0)}$ reduced. Let $h_f : V_f \to V_f$ be the monodromy transformation of the singularity f acting on its Milnor fiber V_f. The zeta function of the monodromy h_f is defined to be

$$\varsigma_f(T) := \prod_{i \geqslant 0} \left[\det(\mathrm{id} - T \cdot (h_f)_* |_{H_i(V_f; \mathbb{C})}) \right]^{(-1)^{i+1}}.$$

A result of Campillo–Delgado–Gusein-Zade [8, Thm. 1] allows us to prove:

Theorem 5 *Let $k = \mathbb{C}$. Then, for every $\mathcal{O} = \mathcal{O}_{(\mathbb{C}^2, 0)}/(f)$, with $f \in \mathcal{O}_{(\mathbb{C}^2, 0)}$ reduced, and for every $S = S(\mathcal{O})$, we have*

$$\varsigma_f(T) = \mathcal{Z}(T_1, \ldots, T_d, U, S) \big|_{\substack{T_1 = \cdots = T_d = T. \\ U = 1}}$$

In [6] Zúñiga introduced a Dirichlet series $Z(\mathrm{Ca}(X), T)$ associated to the effective Cartier divisors on an algebraic curve X defined over $k = \mathbb{F}_q$, which admits an Euler product of the form

$$Z(\mathrm{Ca}(X), T) = \prod_{P \in X} Z_{\mathrm{Ca}(X)}(T, q, \mathcal{O}_{P, X}),$$

with $Z_{\mathrm{Ca}(X)}(T, q, \mathcal{O}_{P, X}) := \sum_{I \subseteq \mathcal{O}_{P, X}} T^{\dim_k(\mathcal{O}_{P, X}/I)}$, where I runs through all the principal ideals of $\mathcal{O}_{P, X}$. In addition, $Z_{\mathrm{Ca}(X)}(T, q, \mathcal{O}_{P, X}) = Z(T, \mathcal{O}_{P, X})$; see Sect. 2.

Observe that this zeta function is nothing but the zeta function $Z(\mathcal{O}, \mathcal{O}, T)$ appearing as a local factor in the Stöhr zeta function, cf. Sect. 1.

Remark 6 In the category of $\mathbb{F}_q$-algebraic varieties, $[\cdot]$ specializes to the counting rational points additive invariant $\#(\cdot)$. We write $\#(Z(T_1, \ldots, T_d, \mathcal{O}))$ for the rational function obtained by specializing $[\cdot]$ to $\#(\cdot)$. From a computational point of view, $\#(Z(T_1, \ldots, T_d, \mathcal{O}))$ is obtained from $Z(T_1, \ldots, T_d, \mathcal{O})$ by replacing $\mathbb{L}$ by q.

Theorem 7 *Let $k = \mathbb{F}_q$ and let $\mathcal{Z}(T_1, \ldots, T_d, U, S)$ be the universal zeta function for S. Moreover, let X be an algebraic curve defined over k, and let $\mathcal{O}_{P, X}$ be the (complete) local ring of X at a singular point P. Assume that k is big enough for X and that $S = S(\mathcal{O}_{P, X})$. Then,*

(i) *for any* $\mathcal{O} = \mathcal{O}_{(\mathbb{C}^2,0)}/(f)$, *with* $f \in \mathcal{O}_{(\mathbb{C}^2,0)}$ *reduced, and* $S = S(\mathcal{O})$,

$$Z_{\mathrm{Ca}(X)}(q^{-1}T, q, \mathcal{O}_{P,X}) = \#\left(Z(T_1, \ldots, T_d, \mathcal{O}_{P,X})\right)$$

$$= \mathcal{Z}(T_1, \ldots, T_d, U, S) \mid_{\substack{T_1 = \cdots = T_d = T. \\ U = q}}$$

In particular, $Z_{\mathrm{Ca}(X)}(q^{-1}T, q, \mathcal{O}_{P,X})$ *depends only on* S. *Moreover, if* X *is plane, then* $Z_{\mathrm{Ca}(X)}(q^{-1}T, q, \mathcal{O}_{P,X})$ *is a complete invariant of the equisingularity class of* $\mathcal{O}_{P,X}$;
(ii) *for any* $\mathcal{O} = \mathcal{O}_{(\mathbb{C}^2,0)}/(f)$, *with* $f \in \mathcal{O}_{(\mathbb{C}^2,0)}$, *it holds that*

$$Z_{\mathrm{Ca}(X)}(q^{-1}T, q, \mathcal{O}_{P,X}) \mid_{q \to 1} = \varsigma_f(T).$$

4 Some Connections Between Local and Global Zeta Functions

For a smooth algebraic variety Y defined over a field k, M. Kapranov defined a zeta function as the formal power series in an indeterminate u

$$\zeta_{\mathrm{mot},Y}(u) = \sum_{n=0}^{\infty} [Y^{(n)}] u^n \in K_0(\mathrm{Var}_k)[\![u]\!],$$

where $Y^{(n)}$ stands for the n-fold symmetric product of Y (cf. [23, Sect. 1]). (For instance, if $k = \mathbb{F}_q$, then one obtains the usual Hasse–Weil zeta function of Y, cf. Remark 6). When Y is a curve, Baldassarri, Deninger and Naumann introduced in [24] a two-variable version of the Kapranov zeta function, namely

$$Z_{\mathrm{mot},Y}(t, u) = \sum_{n,\nu \geq 0} [\mathrm{Pic}_\nu^n] \frac{u^\nu - 1}{u - 1} t^n \in K_0(\mathrm{Var}_k)[\![u, t]\!],$$

where the algebraic k-scheme $\mathrm{Pic}_\nu^n = \mathrm{Pic}_{\geq \nu}^n \setminus \mathrm{Pic}_{\geq \nu+1}^n$ (with $\mathrm{Pic}_{\geq \nu}^n$ being the closed subvariety—in the Picard variety of degree n line bundles on Y—of line bundles $\mathcal{L}$ with $h^0(\mathcal{L}) \geq \nu$) defines a class in $K_0(\mathrm{Var}_k)$.

The connections between the universal zeta function and the motivic zeta functions of Kapranov and Baldassarri–Deninger–Naumann are being currently investigated by A. Melle, W. Zúñiga and the author; we believe that the zeta functions discussed in the previous sections are factors of motivic zeta functions of Baldassarri–Deninger–Naumann type for singular curves (as mentioned before, this is known when $k = \mathbb{F}_q$). In order to give some evidence supporting this belief, this note will be finished by stating the relation between local and global zeta functions in a particular situation.

The context will be the one of a *modulus*: following Serre [22], let k be an algebraically closed field, and let C be an irreducible, non-singular, complete algebraic curve defined over k. If F is a finite subset of C, a modulus $\mathfrak{m}$ supported on F is defined to be the assignment of an integer $n_P > 0$ for each point $P \in F$; this is sometimes identified with the effective divisor $\sum_P n_P P$.

It is possible to attach a curve to $\mathfrak{m}$ starting from C, essentially by "placing" the points in F all together into one; see again [22]. The resulting singular curve $C_\mathfrak{m}$ has then this point as its only singularity. It holds the following

Theorem 8 *Let $C_\mathfrak{m}$ be a curve arising from a modulus $\mathfrak{m}$ supported on a finite set of points of a curve C as above, and let P be its only singular point. Furthermore, let $\pi: \widetilde{C_\mathfrak{m}} \to C_\mathfrak{m}$ be the normalization morphism. Then,*

$$
Z_{\mathrm{mot},C_\mathfrak{m}}(\mathbb{L}^{-1}T, \mathbb{L}) = Z_{\mathrm{mot},\widetilde{C_\mathfrak{m}}}(\mathbb{L}^{-1}T, \mathbb{L}) \prod_{i=1}^{\sharp(\pi^{-1}(P))} (1 - \mathbb{L}^{-1}T) \cdot Z(T, \mathcal{O}_P).
$$

The proof of this statement will appear in a forthcoming paper.

References

1. F.K. Schmidt, Analytische zahlentheorie in körpern der charakteristik p. Math. Z. **33**, 1–32 (1931)
2. K.O. Stöhr, On the poles of regular differentials of singular curves, Bol. Soc. Brasil. Mat. (N.S.) **24**(1), 105–136 (1993)
3. K.O. Stöhr, Local and global zeta-functions of singular algebraic curves. J. Number Theory **71**(2), 172–202 (1998)
4. V.M. Galkin, Zeta functions of some one-dimensional rings. Izv. Akad. Nauk. SSSR Ser. Mat. **37**, 3–19 (1973)
5. B. Green, Functional equations for zeta-functions of non-Gorenstein orders in global fields. Manuscripta Math. **64**, 485–502 (1989)
6. W.A. Zúñiga-Galindo, Zeta functions and Cartier divisors on singular curves over finite fields. Manuscripta Math. **94**, 75–88 (1997)
7. A. Campillo, F. Delgado, K. Kiyek, Gorenstein property and symmetry for one-dimensional local Cohen-Macaulay rings. Manuscripta Math. **83**(3–4), 405–423 (1994)
8. A. Campillo, F. Delgado, S. Gusein-Zade, On the monodromy of a plane curve singularity and the Poincaré series of its ring of functions. (Russian) Funktsional. Anal. i Prilozhen. **33**(1), 66–68 (1999); translation in Funct. Anal. Appl. **33**(1), 56–57 (1999)
9. A. Campillo, F. Delgado, S. Gusein-Zade, Poincaré series of curves on rational surface singularities. Comment. Math. Helv. **80**(1), 95–102 (2005)
10. A. Campillo, F. Delgado, S. Gusein-Zade, The Alexander polynomial of a plane curve singularity, and the ring of functions on the curve. (Russian) Uspekhi Mat. Nauk **327**(3), 157–158 (1999); translation in Russian Math. Surveys **54**(3), 634–635 (1999)
11. A. Campillo, F. Delgado, S. Gusein-Zade, Integrals with respect to the Euler characteristic over spaces of functions, and the Alexander polynomial. (Russian) Tr. Mat. Inst. Steklova **238**, 144–157 (2002); Monodromiya v Zadachakh Algebr. Geom. i Differ. Uravn. translation in Proc. Steklov Inst. Math. **238**(3), 134–147 (2002)
12. A. Campillo, F. Delgado, S. Gusein-Zade, The Alexander polynomial of a plane curve singularity and integrals with respect to the Euler characteristic. Int. J. Math. **14**(1), 47–54 (2003)

13. A. Campillo, F. Delgado, S. Gusein-Zade, The Alexander polynomial of a plane curve singularity via the ring of functions on it. Duke Math. J. **117**(1), 125–156 (2003)
14. A. Campillo, F. Delgado, S. Gusein-Zade, On the zeta functions of a meromorphic germ in two variables, in *Geometry, Topology, and Mathematical Physics*, American mathematical society translations: series 2 (Amer. Math. Soc., Providence, RI, 2004). 212
15. O. Viro, Some integral calculus based on Euler characteristic, in *Topology and Geometry-Rohlin Seminar*, Lecture notes in math, ed. by O.Ya. Viro (Springer, Heidelberg, 1988). 1346
16. A. Campillo, F. Delgado, S. Gusein-Zade, Multi-index filtrations and generalized Poincaré series. Monatsh. Math. **150**(3), 193–209 (2007)
17. J.J. Moyano-Fernández, Poincaré series associated with curves defined over perfect fields. Ph.D. thesis, Universidad de Valladolid, 2008
18. F. Delgado, J.J. Moyano-Fernández, On the relation between the generalized Poincaré series and the Stöhr zeta function. Proc. Am. Math. Soc. **137**(1), 51–59 (2009)
19. K.O. Stöhr, Multi-variable Poincaré series of algebraic curve singularities over finite fields. Math. Z. **262**(4), 849–866 (2009)
20. J.J. Mira, K.O. Stöhr, On Poincaré series of singularities of algebraic curves over finite fields. Manuscripta Math. **147**(3–4), 527–546 (2015)
21. J.J. Moyano-Fernández, W.A. Zúñiga-Galindo, Motivic zeta functions for curve singularities. Nagoya Math. J. **198**, 47–75 (2010)
22. J.-P. Serre, *Algebraic Groups and Class Fields. Graduate Texts in Mathematics*, vol. 117 (Springer, Heidelberg, 1988)
23. M. Kapranov, The elliptic curve in the *S*-duality theory and Eisenstein series for Kac-Moody groups. arXiv:math/0001005 [math.AG]. preprint
24. F. Baldassarri, C. Deninger, N. Naumann, A motivic version of Pellikaan's two variable zeta function, in *Diophantine Geometry*, Norm edn., CRM Series, vol. 4 (Pisa, 2007), pp. 35–43

Algebraic Volumes of Divisors

Carsten Bornträger and Matthias Nickel

Abstract We prove the following result: for every totally real Galois number field K there exists a smooth projective variety X and a divisor D on X such that $vol_X(D)$ is a primitive element of K.

1 Introduction

The volume of a Cartier divisor D on a projective complex variety X measures the asymptotic rate of growth of global sections of its multiples. If $\dim(X) = d$, then

$$\mathrm{vol}_X(D) = \limsup_{n \to \infty} \frac{h^0(\mathcal{O}_X(nD))}{n^d/d!} \, .$$

For nef divisors the asymptotic Riemann–Roch theorem yields $\mathrm{vol}_X(D) = (D^d)$.

The volume was first used implicitly by Cutkosky [1] to study the existence of Zariski decomposition on higher dimensional varieties and has evolved into a fundamental invariant of line bundles in projective geometry. In [1] Cutkosky shows that there is an effective divisor on a certain threefold which has irrational volume, hence can not have a Zariski decomposition even after a birational modification.

The volume function itself enjoys many interesting formal properties: it depends only on the numerical equivalence class of the divisor and can be extended to a continuous function on the real Néron–Severi space $N^1_{\mathbb{R}}(X)$.

We will be primarily interested in the set of volumes

C. Bornträger (✉)
Phillips-Universität Marburg, Institut für Mathematik, Hans-Meerwein-Straße,
35032 Marburg, Germany
e-mail: borntraegerc@mathematik.uni-marburg.de

M. Nickel
Goethe-Universität, Institut für Mathematik, Robert-Mayer-Str. 6–8,
60325 Frankfurt am Main, Germany
e-mail: nickel@math.uni-frankfurt.de

© Springer Nature Switzerland AG 2018
M. Alberich-Carramiñana et al. (eds.), *Extended Abstracts February 2016*,
Trends in Mathematics 9, https://doi.org/10.1007/978-3-030-00027-1_13

$\mathbb{V} := \{a \in \mathbb{R}_+ \mid a = \mathrm{vol}_X(D) \text{ for some pair } (X, D), D \text{ integral Cartier divisor}\},$

which is known to contain $\mathbb{Q}_+$ and is a multiplicative semigroup by the Künneth formula [2, Rem. 2.42].

The volume function is locally piecewise polynomial with rational coefficients on surfaces [3]. Another case of this phenomenon is the following: let X be a $\mathbb{Q}$-factorial projective variety and let $N^1_{\mathbb{R}}(X)$ be spanned by finitely many numerical equivalence classes of effective $\mathbb{Q}$-divisors who satisfy that their Cox ring is finitely generated, that an ample divisor is contained in the cone generated by them and that every divisor class in the interior of this cone contains only divisors whose ring of sections is finitely generated. Then the volume function is locally piecewise polynomial with rational coefficients on the cone generated by these divisors [4]. On the other hand it can can be irrational or even transcendent in the absence of finite generatedness properties [3, Subsect. 3.3], [5, Sect. 3]. Furthermore in [5] the authors verify that there are only countably many volume functions for all irreducible projective varieties, in particular, $\mathbb{V}$ is countable. They also show the existence of a fourfold where the volume function is given by a transcendental formula on an open subset of the big cone $\mathrm{Big}(X)_{\mathbb{R}}$, which illustrates that the behaviour of $\mathbb{V}$ is rather mysterious.

We intend to show that $\mathbb{V}$ contains a large class of positive real algebraic numbers as follows.

Theorem 1 *For every totally real Galois number field K there exists a smooth projective variety X and a divisor D on X such that $\mathrm{vol}_X(D)$ is a primitive element of K.*

There is also a strong connection to Okounkov bodies: in [6] it is shown that

$$\mathrm{vol}_X(D) = d! \int_{\Delta_{Y_\bullet}(D)} 1 \, d\mu,$$

where $\Delta_{Y_\bullet}(D)$ is the Okounkov body associated to D and an admissible complete flag $Y_\bullet$, and μ is the Lebesgue measure on $\mathbb{R}^d$.

2 Preliminaries

2.1 Volumes on Projective Bundles

Let V be an irreducible projective variety of dimension v, and $A_0, \ldots, A_r$ be Cartier divisors on V. We set $\mathcal{E} = \mathcal{O}_V(A_0) \oplus \cdots \oplus \mathcal{O}_V(A_r)$ and consider the projective bundle $X = \mathbb{P}(\mathcal{E})$, which is an irreducible projective variety of dimension $d = v + r$.

One then has the following Lemma, which is well known.

Lemma 2

$$\mathrm{vol}_X\,(\mathcal{O}_X(1)) = \frac{d!}{v!} \int_{\substack{\lambda_0+\cdots+\lambda_r=1 \\ \lambda_i \geq 0}} \mathrm{vol}_V(\lambda_0 A_0 + \cdots + \lambda_r A_r)\,\mathrm{d}\lambda_1 \cdots \mathrm{d}\lambda_r.$$

Remark 3 Let E be an elliptic curve without complex multiplication and take $V = E \times E$. The nef cone of V is equal to its effective cone and it is circular. Taking suitable A_0 ample and A_1 not nef, Cutkosky utilizes the geometry of $\mathrm{Nef}(V)$ to deduce that $\mathrm{vol}_X(\mathcal{O}_X(1))$ is a quadratic irrationality.

2.2 Abelian Varieties

Let (A, L_0) be a polarized d-dimensional complex abelian variety, then each $L \in N^1(A)$ induces a homomorphism $\phi_L\colon A \to \hat{A}$, where $\hat{A}$ is the dual abelian variety of A. The map ϕ_{L_0} is an isogeny, and we obtain an isomorphism of $\mathbb{Q}$-vector spaces

$$\begin{aligned} \varphi\colon N^1_{\mathbb{Q}}(A) &\to \mathrm{End}^s_{\mathbb{Q}}(A) \\ L &\mapsto \phi_{L_0}^{-1}\phi_L, \end{aligned} \tag{1}$$

where $\mathrm{End}^s_{\mathbb{Q}}(X)$ denotes the subspace of $\mathrm{End}_{\mathbb{Q}}(A)$ fixed by the Rosati involution with respect to the polarization L_0; see Birkenhake–Lange [7, Prop. 5.2.1].

Proposition 5.2.3 from Birkenhake–Lange [7] shows that the characteristic polynomial $P^a_{f_L}$ of the analytic representation f_L of $\phi_{L_0}^{-1}\phi_L \in \mathrm{End}^s_{\mathbb{Q}}(X)$ satisfies

$$P^a_{f_L}(t) = \frac{(tL_0 - L)^d}{d_0 d!},$$

where d_0 denotes the degree of the polarization L_0.

From now on we concentrate on abelian varieties with real multiplication.

Lemma 4 *Let K be a totally real number field of degree d over $\mathbb{Q}$ with primitive element α. Then there exists a d-dimensional polarized simple abelian variety (A, L_0) and a line bundle L on A such that the volume function $\mathrm{vol}_A(tL_0 - L)$ restricted to the nef cone of A is given by a rational multiple of the minimal polynomial of α.*

Proof It is possible to construct a simple, polarized, abelian variety with $\mathrm{End}_{\mathbb{Q}}(A) = K$; see [7, Prop. 9.2.1]. We also have $\mathrm{End}^s_{\mathbb{Q}}(A) = \mathrm{End}_{\mathbb{Q}}(A)$ since K is totally real; see [7, Prop. 5.5.7]. Take (A, L_0) with this property and let L be $\varphi^{-1}(\alpha)$. The minimal polynomial of α is equal to its characteristic polynomial because they both have degree d. By the above discussion and the Riemann–Roch theorem the lemma is proven.

In the case of abelian varieties the boundary of the nef cone has the following property; see also Lazarsfeld [8, Cor. 1.5.18].

Lemma 5 *Let A be an abelian variety of dimension d and $\delta \in N^1_{\mathbb{R}}(A)$ a numerical equivalence class on A which lies in the boundary of the nef cone. Then, $(\delta^d) = 0$.*

Furthermore abelian varieties have the useful property that the nef cone and the pseudoeffective cone coincide. This is even true in a more general setting; see [8, Ex. 1.4.7].

Lemma 6 *Let V be a complete variety with a connected algebraic group acting transitively on it. Then, every effective divisor on V is nef.*

3 Algebraic Volumes of Divisors

We now proceed with the proof of the main theorem.

(*Proof of Theorem* 1) We use Lemma 4 to find a polarized abelian variety (A, L_0) with the prescribed properties. Using the isomorphism (1) we take L to correspond to a primitive element α of K. We choose $t_0 \in \mathbb{N}$ large enough so that $t_0 L_0 - L$ is ample.

Considering Cutkosky's construction with $r = 1$, $A_0 = -L$, and $A_1 = t_0 L_0 - L$ so that $X = \mathbb{P}(\mathcal{O}_A(-L) \oplus \mathcal{O}_A(t_0 L_0 - L))$, we will compute $\mathrm{vol}_X(\mathcal{O}_X(1))$.

Lemma 2 yields that $\mathrm{vol}_X(\mathcal{O}_X(1))$ is a rational multiple of the integral

$$\int_0^{t_0} \mathrm{vol}_A(t L_0 - L) \, dt \,,$$

while Lemma 4 shows that the latter is a rational multiple of

$$\int_\beta^{t_0} m_\alpha(t) \, dt$$

with m_α the minimal polynomial of α over $\mathbb{Q}$, and β the largest root of m_α. This number is not necessarily a primitive element of K, but it can be shown that primitivity holds for a general choice of α.

Question 7 *Is it possible to extend this method to show that every nonnegative (totally) real algebraic number appears as $\mathrm{vol}_X(D)$ for some pair (X, D)?*

References

1. S.D. Cutkosky, Zariski decomposition of divisors on algebraic varieties. Duke Math. J. **53**(1), 149–156 (1986)
2. A. Küronya, *Partial Positivity Concepts in Projective Geometry* (Albert-Ludwigs-Universität Freiburg, Habilitationsschrift, 2011)

3. T. Bauer, A. Küronya, T. Szemberg, Zariski chambers, volumes, and stable base loci. J. Reine Angew. Math. **576**, 209–233 (2004)
4. A.-S. Kaloghiros, A. Küronya, V. Lazić, *Finite Generation and Geography of Models, to appear in Minimal Models and Extremal Rays*, Advanced studies in pure mathematics (Mathematical Society of Japan, Tokyo., 2012)
5. A. Küronya, V. Lozovanu, C. Maclean, Volume functions of linear series. Math. Ann. **356**(2), 635–652 (2013)
6. R. Lazarsfeld, M. Mustaţă, Convex bodies associated to linear series, Ann. Sci. Éc. Norm. Supér. (4) **42**(5), 783–835 (2009)
7. C. Birkenhake, H. Lange, *Complex Abelian Varieties*, 2nd edn., Grundlehren der Mathematischen Wissenschaften [Fundamental Principles of Mathematical Sciences], vol. 302 (Springer, Berlin, 2004)
8. R. Lazarsfeld, *Positivity in algebraic geometry. I*, Ergebnisse der Mathematik und ihrer Grenzgebiete. 3. Folge. A Series of Modern Surveys in Mathematics [Results in Mathematics and Related Areas. 3rd Series. A Series of Modern Surveys in Mathematics], vol. 48 (Springer-Verlag, Berlin, 2004). Classical setting: line bundles and linear series

On Hirzebruch Type Inequalities and Applications

Justyna Szpond

Abstract In a series of articles, Hirzebruch studied Kummer covers of complex projective plane branched along arrangements of lines. His studies were aimed at constructing surfaces of general type with interesting invariants, in particular with Chern classes satisfying the equality $c_1^2(X) = 3c_2(X)$. This interest was motivated by one of central results in the theory of surfaces of general type to the effect that there is always the inequality $c_1^2(X) \leq 3c_2(X)$. This fact is known as the Miyaoka–Yau inequality. Hirzebruch proved in passing a number of remarkable inequalities involving invariants of line arrangements. No combinatorial proofs of these inequalities seem to be known. The purpose of this note is to report on these inequalities and put them in the perspective of more recent results in combinatorics.

1 Introduction

By about 1960 the classification program initiated by Federigo Enriques for compact complex algebraic surfaces was completed with results of Kunihiko Kodaira. The attention turned to a finer problem known as the geography of surfaces of general type; see, e.g., [1]. This problem amounts to deciding when a pair (p, q) of integers appears as Chern numbers $c_1^2(X) = p$ and $c_2(X) = q$ of a minimal surface of general type X. Along these lines Antonius Van de Ven [2] proved the restriction $c_1^2(X) \leq 8c_2(X)$. This result was improved by Fedor Bogomolov [3] to $c_1^2(X) \leq 4c_2(X)$ and shortly after improved further by Yoichi Miyaoka [4] and Shing-Tung Yau [5] to the following celebrated result.

Theorem 1 (Miyaoka–Yau inequality) *Let X be a minimal surface of general type then,*

$$c_1^2(X) \leq 3c_2(X). \tag{1}$$

J. Szpond (✉)
Department of Mathematics, Pedagogical University of Cracow, Podchorążych 2,
30-084 Kraków, Poland
e-mail: szpond@up.krakow.pl

Moreover Yau's proof implied that whenever there is the equality in (1), X is a ball quotient.

Over the years this inequality has been extensively studied and generalized in various ways. We recall here a mutation convenient for subsequent considerations.

Theorem 2 (Logarithmic Miyaoka–Yau inequality) *Let X be a smooth projective surface of non-negative Kodaira dimension and let D be a simple normal crossing divisor on X. Then, $c_1^2(\Omega_X^1(\log D)) \leq 3c_2(\Omega_X^1(\log D))$.*

An example of a surface with $c_1^2(X) = 3c_2(X)$ was constructed before by Armand Borel [6], so that it was known that the constant coefficient in the inequality (1) could not have been improved any more. Nevertheless it had been of substantial interest to construct further examples and to investigate into the existence of surfaces X with the ratio $c_1^2(X)/c_2(X)$ in the interval [2, 3]. This question motivated Friedrich Hirzebruch to study Kummer covers of the projective plane branched over arrangements of lines [7, 8].

Definition 3 (*Arrangement of lines*) An *arrangement of lines* $\mathcal{L}$ in the projective plane $\mathbb{P}^2$ is a union of finitely many mutually distinct lines $L_1, \ldots, L_d$.

A point $P \in \mathbb{P}^2$ is a point of multiplicity $k \geq 2$ of $\mathcal{L}$ if there are exactly k lines from $\mathcal{L}$ intersecting at P. The number of points of multiplicity k for $\mathcal{L}$ is denoted by $t_k(\mathcal{L})$ or simply t_k if $\mathcal{L}$ is understood.

In order to decide if constructed surfaces are of general type, Hirzebruch proved in passing the following result which is the main object of our interest in this note.

Theorem 4 (Hirzebruch inequalities) *Let $\mathcal{L}$ be an arrangement of lines such that $t_d = t_{d-1} = 0$ then, $t_2 + t_3 \geq d + \sum_{k \geq 4}(k - 4)t_k$.*

2 Sylvester–Gallai Type Theorems

In this section we pass to the dual setting: given a finite set of points $\mathcal{P} = \{P_1, \ldots, P_s\}$ we consider the set $\mathcal{L}(\mathcal{P})$ of lines determined by pairs of points in $\mathcal{P}$. By a slight abuse of notation we denote by $t_k = t_k(\mathcal{P})$ the number of lines in $\mathcal{L}(\mathcal{P})$ which contain exactly k points from $\mathcal{P}$. With this notation Hirzebruch Inequalities hold for finite sets of points in the complex (and hence also real) projective plane verbatim.

Definition 5 (*A Sylvester–Gallai configuration*) We say that a finite set of points $\mathcal{P}$ is a Sylvester–Gallai configuration (an SG-configuration for short) if for any two points $P, Q \in \mathcal{P}$ there exists a third point $R \in \mathcal{P}$ collinear with P and Q.

In other words, $\mathcal{P}$ is an SG-configuration if $t_2(\mathcal{P}) = 0$.

A considerable amount of research in the real projective geometry and combinatorics has been motivated by the following celebrated result.

Theorem 6 (Sylvester–Gallai) *Let $\mathcal{P} \subset \mathbb{P}^2(\mathbb{R})$ be a finite set of s points. Then,*

(i) either all points are collinear, i.e., $t_s = 1$;
(ii) or there is a line containing exactly two points from $\mathcal{P}$, i.e., $t_2 > 0$.

In other words, the only *real* SG-configurations are contained in a line. There are numerous proofs of the Sylvester–Gallai Theorem. One of them is based on the following inequality obtained by Eberhard Melchior [9].

Theorem 7 (Melchior inequality) *Let $\mathcal{P}$ be a finite set of s points in the real plane such that $t_s = 0$. Then, $t_2 \geq 3 + \sum_{k \geq 3}(k-3)t_k$.*

The Hirzebruch inequality can be viewed as a complex analog of the Melchior inequality. It should be however stressed that whereas Melchior Inequality follows rather simply by counting the topological Euler characteristic of the real projective plane using the partition imposed by the arrangement of lines, Hirzebruch Inequalities are based on much deeper results.

A possible generalization of Theorem 6 to the complex projective plane is the following.

Theorem 8 (A complex SG-type theorem) *Let $\mathcal{P} \subset \mathbb{P}^2(\mathbb{C})$ be a finite set of s points. Then,*

(i) either all points are collinear, i.e., $t_s = 1$;
(ii) or there is a line containing exactly either two or three points from $\mathcal{P}$, i.e., $t_2 > 0$ or $t_3 > 0$.

It is well known that there are non-collinear SG-configurations in the complex plane. The simplest example is provided by the so called Hesse configuration.

Example 9 (*Hesse configuration*) In this example $\mathcal{P}$ is the set of 3-division points on an elliptic curve embedded as a smooth cubic curve into $\mathbb{P}^2$. In a suitable system of coordinates one can take, e.g., the Fermat cubic $x^3 + y^3 + z^3 = 0$. Then, with ε a primitive 3-rd root of 1, the 3-division points are

$$P_1 = (0:1:-1), \; P_2 = (0:1:-\varepsilon), \; P_3 = (0:1:-\varepsilon^2),$$
$$P_4 = (1:0:-1), \; P_5 = (1:0:-\varepsilon), \; P_6 = (1:0:-\varepsilon^2),$$
$$P_7 = (1:-1:0), \; P_8 = (1:-\varepsilon:0), \; P_9 = (1:-\varepsilon^2:0).$$

These points in the Hesse configuration are contained in the union of three lines. For example one can take the coordinate axes: $x = 0$, $y = 0$ and $z = 0$. These lines form a triangle. It was observed by Kelly [10] that there is no SG-configuration contained in three concurrent lines.

Lemma 10 *A complex SG-configuration of points cannot be contained in the union of three concurrent lines.*

Proof Assume to the contrary that $\mathcal{P} = \{P_1, \ldots, P_s\}$ is an SG-configuration contained in three lines L_1, L_2, L_3 meeting in a point Z. Without loss of generality we may assume that these three lines have equations $y = 0$, $y - z = 0$, and $z = 0$, respectively. Hence, $Z = (1 : 0 : 0)$. It is convenient to rename the points in $\mathcal{P}$ according to their position on the lines. Let

$$L_1 \cap (\mathcal{P} \setminus \{Z\}) = \{A_1, \ldots, A_t\}$$
$$L_2 \cap (\mathcal{P} \setminus \{Z\}) = \{B_1, \ldots, B_t\}$$
$$L_3 \cap (\mathcal{P} \setminus \{Z\}) = \{C_1, \ldots, C_t\}.$$

The number of points in each set is the same. Indeed, joining for example the point C_1 with $A_1, \ldots, A_t$ on L_1 defines points $B_1, \ldots, B_t$ on L_2 (thus C_1, A_i, B_i are assumed to be collinear for $i = 1, \ldots, t$). There cannot be any additional points from $\mathcal{P}$ other than $B_1, \ldots, B_t$ on L_2 since the line $C_1 B$ through such an additional point B would contain only 2 points from $\mathcal{P}$. So the number of points from $\mathcal{P}$ on L_1 and L_2 is the same. Analogous argument starting with the point A_1 in the place of C_1 shows that the number of points from $\mathcal{P}$ on L_2 and L_3 is the same.

We can additionally specify $A_1 = (0 : 0 : 1)$, $B_1 = (0 : 1 : 1)$ and consequently $C_1 = (0 : 1 : 0)$. Note that the line $A_i B_1$, where $A_i = (a_i : 0 : 1)$, is given by the equation $x + a_i y - a_i z = 0$ and it intersects the line L_3 in the point $C_i = (-a_i : 1 : 1)$. Further, the line $B_j C_i$, where $B_j = (a_j : 1 : 1)$, is given by $x + a_i y - (a_i + a_j)z = 0$ and it intersects the line L_1 in the point $A_k = (a_i + a_j : 0 : 1)$. Hence the set $\{A_1, \ldots, A_t\}$ considered as a subset of the affine line $L_1 \setminus \{Z\} \simeq \mathbb{C}$ is closed under the addition which is not possible. This contradiction ends the proof. $\square$

The observation that all real SG-configurations are contained in a line and the existence of complex planar SG-configurations motivated Jean-Pierre Serre to ask in [11] a question about higher dimensional SG-configurations. This question was answered by Leroy Milton Kelly 20 years later, see [10]. We present its proof as it applies in a very nice way the Hirzebruch Inequality.

Theorem 11 (A complex Sylvester–Gallai theorem) *Let* $\mathcal{P} = \{P_1, \ldots, P_s\}$ *be a finite set of points in* $\mathbb{P}^n(\mathbb{C})$ *with* $n \geq 3$. *Then,*

(i) either $\mathcal{P}$ *is contained in a plane* $H = \mathbb{P}^2(\mathbb{C}) \subset \mathbb{P}^n(\mathbb{C})$;
(ii) or there exists a line passing through exactly two points from the set $\mathcal{P}$.

Proof Assume that there is an SG-configuration $\mathcal{P}$ not contained in a plane. Since a central projection of an SG-configuration to a hyperplane is again an SG-configuration, we can assume that $\mathcal{P}$ is contained in $\mathbb{P}^3(\mathbb{C})$ and that it is not contained in any plane in $\mathbb{P}^3(\mathbb{C})$. Let $Z \in \mathcal{P}$ and H be a point and a plane such that $Z \notin H$. Let $\mathcal{P}'$ be the image of $\mathcal{P}$ under the projection to H centered in Z. If $\mathcal{P}'$ is contained in a line, then $\mathcal{P}$ was contained in the plane and we are done. Otherwise, since $t_2(\mathcal{P}') = 0$ Hirzebruch inequality implies that there is a line L in H containing exactly 3 points A, B, C from $\mathcal{P}'$. Let G be the plane generated by Z and L. Then $\mathcal{P}'' = G \cap \mathcal{P}$ is an SG-configuration. Indeed, any line generated by two points in $\mathcal{P}''$ is contained in G

and since it is also generated by (the same) two points from $\mathcal{P}$, it must contain a third point from $\mathcal{P}$, which also lies in G and in $\mathcal{P}''$. On the other hand $\mathcal{P}''$ is contained in the union of three lines, namely those generated by Z and points A, B and C respectively. This contradicts however Lemma 10 and we are done. $\qquad\square$

Recently Theorem 11 has been revised by Elkies–Pretorius–Swanepoel [12]. In particular, they presented the following result generalizing Theorems 6 and 11 to the quaternions.

Theorem 12 (A quaternion SG-type theorem) *Let $\mathcal{P} = \{P_1, \ldots, P_s\}$ be a finite set of points in $\mathbb{P}^n(\mathbb{H})$ with $n \geq 3$. Then,*

(i) either $\mathcal{P}$ is contained in a space $H = \mathbb{P}^3(\mathbb{H}) \subset \mathbb{P}^n(\mathbb{H})$;
(ii) or there exists a line passing through exactly two points from the set $\mathcal{P}$.

Interestingly, it is not known if this result is sharp, i.e., no example a quaternion SG-configuration generating a 3-dimensional space is known.

Problem 13 Does there exist a configuration points in $\mathbb{P}^3(\mathbb{H})$ such that a line through every pair of points in the configuration contains a third point from the configuration?

Theorem 11 is reproved in [12] in passing without using the Hirzebruch inequality. Yet another proof, also avoiding the inequality has been recently announced in [13]. Nevertheless it seems that Kelly's original proof based on the Inequality of Hirzebruch remains the most elegant one, at least from the algebraic point of view.

3   De Bruijn–Erdös Theorem

In this section we show how the De Bruijn–Erdös theorem (see [14]) for complex lines follows quickly from the Hirzebruch inequality, i.e., the logarithmic Miyaoka–Yau inequality Theorem 2.

We begin by recalling the following simple but useful combinatorial equality.

Lemma 14 (A combinatorial equation) *Let $\mathcal{L}$ be an arrangement of d lines defined over an arbitrary field. Then, $\binom{d}{2} = \sum_{k \geq 2} t_k \binom{k}{2}$.*

Theorem 15 *Given a configuration $L = L_1 + \cdots + L_d$ of d complex lines, $\sum_{k \geq 2} t_k \geq d$ unless the lines belong to the same pencil ($t_d = 1$).*

Proof Assume that the configuration does not form a pencil. Let $f : X \to \mathbb{P}^2$ be the blow up of all points with multiplicity $m_P \geq 3$. The exceptional divisor of f over the point P is denoted by E_P. Let $D \subset X$ be the reduced total transform of the configuration. Then,

$$D = f^* L - \sum_{P : m_P \geq 3} (m_P - 1) E_P$$

is a simple normal crossing divisor with

$$e(D) = 2(d + \sum_{k \geq 3} t_k) - (t_2 + \sum_{k \geq 3} kt_k).$$

For X we have

$$e(X) = 3 + \sum_{k \geq 3} t_k \quad \text{and} \quad K_X = -3H + \sum_{P:m_P \geq 3} E_P,$$

where H is the pull back of the class of a line under f. From Theorem 2 we have

$$((d - 3)H - \sum_{P:\, m_P \geq 3} (m_P - 2)E_P)^2 \leq 3(3 + \sum_{k \geq 3} t_k - 2(d + \sum_{k \geq 3} t_k) + t_2 + \sum_{k \geq 3} kt_k).$$

Now, using the equality in Lemma 14 and after elementary calculations we arrive to the inequality asserted in the Theorem. □

Acknowledgements This work has been partially supported by National Science Centre, Poland, grant 2014/15/B/ST1/02197.

References

1. B. Hunt, Complex manifold geography in dimension 2 and 3. J. Differ. Geom. **30**, 51–153 (1989)
2. A. van de Ven, On the Chern numbers of certain complex and almost complex manifolds. Proc. Nat. Acad. Sci. U.S.A. **55**, 1624–1627 (1966)
3. F.A. Bogomolov, Holomorphic tensors and vector bundles on projective manifolds. Math. USSR-Izv. **13**(3), 499–555 (1979)
4. Y. Miyaoka, On the Chern numbers of surfaces of general type. Invent. Math. **42**, 225–237 (1977)
5. S.-T. Yau, Calabi's conjecture and some new results in algebraic geometry. Proc. Nat. Acad. Sci. U.S.A. **74**(5), 1798–1799 (1977)
6. A. Borel, Compact Clifford-Klein forms of symmetric spaces. Topology **2**, 111–122 (1963)
7. F. Hirzebruch, Arrangements of lines and algebraic surfaces, in *Arithmetic and Geometry: Papers Dedicated to I.R. Shafarevich on the Occasion of his Sixtieth Birthday*, ed. by I.R. Shafarevich, M. Artin, J.T. Tate (Birkhäuser, Boston, 1983), pp. 113–140
8. F. Hirzebruch, Singularities of algebraic surfaces and characteristic numbers, in *The Lefschetz Centennial Conference*, ed. by D. Sundararaman, S. Gitler, A. Verjovsky (Am. Math. Soc, Providence, 1986), pp. 141–155
9. E. Melchior, Über Vielseite der projektiven Ebene. Deutsche Math. **5**, 461–475 (1940)
10. L.M. Kelly, A resolution of the Sylvester-Gallai problem of J.-P. Serre. Discret. Comput. Geom. **1**(2), 101–104 (1986)
11. J.-P. Serre, Problem 5359. Am. Math. Mon. **73**, 89 (1966)
12. N. Elkies, L.M. Pretorius, K.J. Swanepoel, Sylvester-Gallai theorems for complex numbers and quaternions. Discret. Comput. Geom. **35**(3), 361–373 (2006)
13. Z. Dvir, S. Saraf, A. Wigderson, Improved rank bounds for design matrices and a new proof of Kelly's theorem, Forum Math. Sigma **2**(e4), 24 (2014)
14. N.G. de Bruijn, P. Erdös, On a combinatorial problem. Indagationes Math. **10**, 421–423 (1948)

On the Completion of Normal Toric Schemes Over Rank One Valuation Rings

Alejandro Soto

Abstract A well known theorem of Sumihiro states that every normal toric variety can be completed equivariantly. Using Zariski–Riemann spaces, we have generalized this result to the setting of normal toric schemes over valuation rings of rank one. The later generalize in a natural way the notion of toric varieties and give a general framework for the study of toric degenerations.

1 Introduction

Given a quasi-compact quasi-separated scheme $\mathcal{S}$, a well known theorem of Nagata says that for any separated $\mathcal{S}$-scheme $\mathcal{Y} \to \mathcal{S}$ of finite type, there exists a proper S-scheme $\mathcal{Y}' \to S$ and an open immersion $\mathcal{Y} \hookrightarrow \mathcal{Y}'$ over $\mathcal{S}$.

In [1], Sumihiro proved an equivariant version of this theorem for normal varieties defined over a field with an action of a connected linear group. Later, in [2] he generalized this result to the relative situation over a normal Noetherian base. These results imply, in particular, an equivariant completion for normal toric varieties over a field and over a discrete valuation ring.

In this report, we present a generalization of this result to the setting of toric schemes over arbitraty valuation rings of rank one. One of the main technical difficulties relies in the fact that these schemes are no longer Noetherian and therefore, many standard results of algebraic geometry cannot be used immediately. For a more complete treatment, see [3, 4].

For this note, we will use the following notation:

- (K, v) is a rank one valued field, i.e., $v \colon K^{\times} \to \mathbb{R}$. With $K^{\circ} := \{x \in K^{\times} \mid v(x) \geq 0\} \cup \{0\}$ the corresponding valuation ring and $K^{\circ\circ} := \{x \in K^{\times} \mid v(x) > 0\} \cup \{0\}$ its maximal ideal. We denote by $\widetilde{K} := K^{\circ}/K^{\circ\circ}$ the residue field and by $\Gamma := v(K^{\times})$ the value group.

A. Soto (✉)
University of Tübingen, Tübingen, Germany
e-mail: alejandro.soto@uni-tuebingen.de

© Springer Nature Switzerland AG 2018
M. Alberich-Carramiñana et al. (eds.), *Extended Abstracts February 2016*,
Trends in Mathematics 9, https://doi.org/10.1007/978-3-030-00027-1_15

- $\mathrm{Spec}(K^\circ) = \{\eta, s\}$, with s and η being the closed and generic point, respectively. For any scheme $\mathscr{Y}$ over K°, we will denote by $\mathscr{Y}_\eta$ the generic fiber $\mathscr{Y} \times_{K^\circ} \mathrm{Spec}(K)$ and by $\mathscr{Y}_s$ the special fiber $\mathscr{Y} \times_{K^\circ} \mathrm{Spec}(\widetilde{K})$.
- Let $\mathbb{T} = \mathrm{Spec}(K^\circ[M])$ be the split torus over K° with character lattice $M \simeq \mathbb{Z}^n$ and generic fiber $T := \mathbb{T}_\eta = \mathrm{Spec}(K[M])$. We denote by $N = \mathrm{Hom}(M, \mathbb{Z})$ the dual lattice of M and by $M_G := M \otimes_{\mathbb{Z}} G$ the base change to G, where $G \subset \mathbb{R}$ is an additive subgroup.

2 What Are $\mathbb{T}$-Toric Schemes over a Valuation Ring?

As in the case of toric varieties over a field or a dvr, the following definition makes emphasis on the multiplicative action of a split torus, contained as an open dense subset, which can be extended to the whole scheme. This leads to a rich family of objects, as we will see later, which is not obvious from the definition if the valuation is not discrete.

Definition 1 A $\mathbb{T}$-toric scheme is an integral separated flat scheme $\mathscr{Y}$ of finite type over K° such that $\mathscr{Y}_\eta$ contains T as an open dense subset and the translation action $T \times_K T \to T$ extends to an algebraic action $\mathbb{T} \times_{K^\circ} \mathscr{Y} \to \mathscr{Y}$ over K°.

Note that if the valuation is trivial the special fiber is empty, hence we recover the usual definition of toric varieties over a field. In this note, we restrict ourselves to the case of toric schemes of finite presentation, however they can be defined in greater generality, i.e. not necessarily of finite type; see Gubler [5].

We remark that in the definition they are not required to be normal. The reason is that there are plenty of non-normal toric schemes which also admit a very nice combinatorial description; see Gubler [5, Sect. 9].

2.1 Classification of Normal $\mathbb{T}$-toric Schemes

Normal toric varieties are classified in terms of strictly convex rational fans. In a similar way, we can classify normal $\mathbb{T}$-toric varieties over K°.

A subset $\sigma \subset N_{\mathbb{R}} \times \mathbb{R}_+$ is called a Γ-admissible cone if it can be written as

$$\sigma = \bigcap_{\text{finite}} \{(w, t) \in N_{\mathbb{R}} \times \mathbb{R}_+ | \langle m_i, w \rangle + c_i t \geq 0\}, \quad m_i \in M, \ c_i \in \Gamma,$$

and contains no subspace of positive dimension. A fan consisting of Γ-admissible cones is called Γ-admissible. We have the following theorem; see Gubler–Soto [3].

Theorem 2 *There is a bijective correspondence between the isomorphism classes of normal $\mathbb{T}$-toric schemes over K° and Γ-admissible fans Σ, whose cones $\sigma \in \Sigma$ satisfy $\sigma \cap (N_{\mathbb{R}} \times \{1\}) \subset N_\Gamma$.*

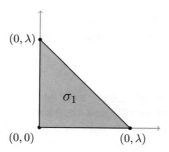

Fig. 1 Polyhedron σ_1

In the affine case, for every normal affine $\mathbb{T}$-toric scheme $\mathscr{Y}$ there exists a Γ-admissible cone σ such that $\mathscr{Y} = \mathrm{Spec}(K[M]^\sigma)$, where

$$K[M]^\sigma = \left\{ \sum_{u \in M} a_u \chi^u \in K[M] \mid \langle u, w \rangle + t v(a_u) \geq 0, \forall (w, t) \in \sigma \right\}.$$

Example 3 Consider the cone $\sigma \subset \mathbb{R}^2 \times \mathbb{R}_+$ generated by the polyhedron σ_1 defined by $\sigma_1 = \mathrm{Conv}\{(0, 0), (0, \lambda), (\lambda, 0)\} \subset \mathbb{R}^2$, with $\lambda = v(a) > 0$ for some $a \in K^\circ$; see Fig. 1.

In this case, we have $K[M]^\sigma = K^\circ[x, y, ax^{-1}y^{-1}] = K^\circ[x, y, z]/(xyz - a)$. We get a $\mathbb{T}$-toric variety $\mathscr{Y}_\sigma = \mathrm{Spec}(K[M]^\sigma)$ whose generic fiber is the toric surface given by $\mathrm{Spec}(K[x, y, z]/(xyz - a))$ and the special fiber is the reduced scheme of finite type over $\widetilde{K}$ given by $\mathrm{Spec}(\widetilde{K}[x, y, z]/(xyz))$. Note that each irreducible component is isomorphic to $\mathbb{A}^2_{\widetilde{K}}$, which clearly is a toric variety over $\widetilde{K}$ with torus $\mathbb{T}_{\widetilde{K}} = \mathrm{Spec}(\widetilde{K}[\mathbb{Z}^2])$.

3 Completion

A natural question to ask is, whether these normal toric schemes over K° admit an equivariant completion or not, as in the case of normal toric varieties over a field. Our main result answer this question affirmatively.

Theorem 4 *Let $\mathscr{Y}$ be a normal $\mathbb{T}$-toric scheme over K°, then there exists a proper $\mathbb{T}$-toric scheme $\mathscr{Y}'$ over K° such that $\mathscr{Y} \hookrightarrow \mathscr{Y}'$ embeds equivariantly as an open dense subset.*

We present an outline of the proof. It follows the same lines as the proof of Nagata's theorem given by Fujiwara–Kato [6]. First, we need the following notion.

Let $\mathcal{U} \subset \mathscr{Y}$ be a $\mathbb{T}$-invariant open subscheme. A blow up of $\mathscr{Y}$ along a closed $\mathbb{T}$-invariant center contained in $\mathscr{Y} \setminus \mathcal{U}$ is called $\mathcal{U}$-*admissible*. Note that the blow up of a $\mathbb{T}$-toric scheme along a closed $\mathbb{T}$-invariant center is a $\mathbb{T}$-toric scheme.

With this, we define the following locally ringed space; see [6, 7]. Keeping the same notation, the $\mathbb{T}$-invariant Zariski–Riemann space associated to the pair $(\mathscr{Y}, \mathcal{U})$ is the locally ringed space

$$\langle \mathscr{Y} \rangle_{\mathcal{U}} := \varprojlim \mathscr{Y}_i,$$

where $\mathscr{Y}_i \to \mathscr{Y}$ are $\mathcal{U}$-admissible blow ups. It has canonically a $\mathbb{T}$-action and one of its main features is the fact that it is quasi-compact.

Recall that $\mathscr{Y}$ can be covered by finitely many affine $\mathbb{T}$-invariant open subschemes $\{\mathscr{U}_i\}$. In particular for each $\mathscr{U}$, there is a proper scheme $\mathscr{U}'$ over K° such that $\mathscr{U} \hookrightarrow \mathscr{U}'$ embeds as an open dense subset. The *partial compactification* of $\mathscr{U}$ relative to $\mathscr{Y}$ is defined as

$$\langle \mathscr{U} \rangle_{\mathrm{pc}}^{\mathscr{Y}} := \langle \mathscr{U} \rangle_{\mathrm{cpt}} \setminus \overline{\langle \mathscr{Y} \rangle_{\mathscr{U}} \setminus \mathscr{U}},$$

where $\langle \mathscr{U} \rangle_{\mathrm{cpt}} := \langle \mathscr{U}' \rangle_{\mathscr{U}}$. It satisfies the following property: given $\mathscr{U}_1 \subset \mathscr{U}_2$, we have $\langle \mathscr{U}_1 \rangle_{\mathrm{pc}}^{\mathscr{Y}} \subset \langle \mathscr{U}_2 \rangle_{\mathrm{pc}}^{\mathscr{Y}}$.

Let $\{\mathscr{U}_i\}$ be a finite affine $\mathbb{T}$-invariant open covering of $\mathscr{Y}$. We define the *Zariski–Riemann compactification* of $\mathscr{Y}$, denoted by $\langle \mathscr{Y} \rangle_{\mathrm{cpt}}$, as the cokernel of the maps p and q coming from the embeddings $\langle \mathscr{U}_i \cap \mathscr{U}_j \rangle_{\mathrm{pc}}^{\mathscr{Y}} \hookrightarrow \langle \mathscr{U}_i \rangle_{\mathrm{pc}}^{\mathscr{Y}}$ and $\langle \mathscr{U}_i \cap \mathscr{U}_j \rangle_{\mathrm{pc}}^{\mathscr{Y}} \hookrightarrow \langle \mathscr{U}_j \rangle_{\mathrm{pc}}^{\mathscr{Y}}$, i.e.,

$$\amalg \langle \mathscr{U}_i \cap \mathscr{U}_j \rangle_{\mathrm{pc}}^{\mathscr{Y}} \overset{p}{\underset{q}{\rightrightarrows}} \amalg \langle \mathscr{U}_i \rangle_{\mathrm{pc}}^{\mathscr{Y}} \to \langle \mathscr{Y} \rangle_{\mathrm{cpt}}.$$

It is endowed canonically with a $\mathbb{T}$-action over K°.

The crucial property that allows us to prove Theorem 4, is that the locally ringed space $\langle \mathscr{Y} \rangle_{\mathrm{cpt}}$ is algebraic, i.e., there exists a scheme $\mathscr{Y}'$ of finite presentation over K° such that $\langle \mathscr{Y}' \rangle_{\mathscr{Y}} = \langle \mathscr{Y} \rangle_{\mathrm{cpt}}$. By construction $\mathscr{Y}'$ is a proper $\mathbb{T}$-toric scheme over K° and $\mathscr{Y} \hookrightarrow \mathscr{Y}'$ embeds as an open dense subset equivariantly.

This proves the equivariant completion in the setting of normal $\mathbb{T}$-toric schemes over K°, extending in this way the result of Sumihiro for toric varieties over a field.

References

1. H. Sumihiro, Equivariant completion. J. Math. Kyoto Univ. **14**, 1–28 (1974)
2. H. Sumihiro, Equivariant completion. II. J. Math. Kyoto Univ. **15**(3), 573–605 (1975)
3. W. Gubler, A. Soto, Classification of normal toric varieties over a valuation ring of rank one. Doc. Math. **20**, 171–198 (2015)
4. A. Soto, Nagata's compactification theorem for normal toric varieties over a valuation ring of rank one. Michigan Math. J. **67**(1), 99–116 (2018)
5. W. Gubler, A guide to tropicalizations, in *Algebraic and Combinatorial Aspects of Tropical Geometry. Contemporary Mathematics*, vol. 589 (Amer. Math. Soc., Providence, 2013), pp. 125–189
6. K. Fujiwara, F. Kato, Foundations of rigid geometry (2014). arXiv:1308.4734v3. preprint
7. M. Temkin, Relative Riemann-Zariski spaces. Isr. J. Math. **185**, 1–42 (2011)

Duality on Value Semigroups

Laura Tozzo

Abstract We consider value semigroup ideals of fractional ideals on certain curve singularities. These satisfy natural axioms defining the class of good semigroup ideals. On this class we develop a purely combinatorial counterpart of the duality on Cohen–Macaulay rings. This is joint work with Philipp Korell and Mathias Schulze.

1 Introduction and Motivation

Let R be a complex algebroid curve with s branches and normalization $R \to \overline{R} \cong \mathbb{C}[[t_1]] \times \cdots \times \mathbb{C}[[t_s]]$. Then, there is a multivaluation map

$$\nu = (\nu_1, \ldots, \nu_s) \colon \overline{R} \to (\mathbb{Z} \cup \{\infty\})^s, \quad x \mapsto (\mathrm{ord}_{t_1}(x), \ldots, \mathrm{ord}_{t_s}(x))$$

which associates to R its *value semigroup* $\Gamma_R = \nu(\{x \in R \mid x \text{ non zero-divisor}\})$

The value semigroup of a curve singularity is an important combinatorial invariant with a long history. It determines the topological type of plane curves. In case R is an irreducible curve Kunz [1] showed that R is Gorenstein if and only if its value semigroup Γ_R is symmetric.

Example 1 Consider the plane algebroid curve $R = \mathbb{C}[[x, y]]/\langle x^7 - y^4 \rangle \cong \mathbb{C}[[t^4, t^7]]$. Then R is Gorenstein and $\Gamma_R = \langle 4, 7 \rangle$ is symmetric.

L. Tozzo (✉)
Dipartimento di Matematica, Universitá degli Studi di Genova, Via Dodecaneso 35,
16154 Genova, Italy
e-mail: tozzo@mathematik.uni-kl.de

L. Tozzo
Department of Mathematics, University of Kaiserslautern,
67663 Kaiserslautern, Germany

© Springer Nature Switzerland AG 2018
M. Alberich-Carramiñana et al. (eds.), *Extended Abstracts February 2016*,
Trends in Mathematics 9, https://doi.org/10.1007/978-3-030-00027-1_16

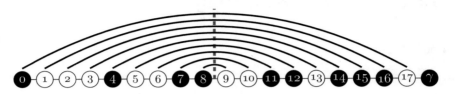

Later Delgado [2] introduced a notion of symmetry in the reducible case, and extended Kunz's result. D'Anna used Delgado's symmetry to define a *canonical semigroup ideal*. Based on this definition, he characterized canonical ideals of R in terms of their value semigroup ideals. More recently, Pol [3] proved a formula for the value semigroup of the dual of a fractional ideal. Our aim is to generalize both the duality results by D'Anna and by Pol.

2   Good Value Semigroups

Including complex algebroid curves as a special case we consider *admissible* rings in the following sense: let R be a one-dimensional semilocal Cohen–Macaulay ring that is analytically reduced, residually rational and has large residue fields (i.e. $|R/\mathfrak{m}| \geq |\{\text{branches of }\widehat{R_\mathfrak{m}}\}|$ for any $\mathfrak{m}$ maximal ideal of R). Value semigroup (ideals) are then defined as follows.

Definition 2 Let R be an admissible ring, and let $\mathfrak{V}_R$ be the set of (discrete) valuation rings of Q_R over R with corresponding valuations $\nu = (\nu_V)_{V \in \mathfrak{V}_R} : Q_R \to (\mathbb{Z} \cup \{\infty\})^{\mathfrak{V}_R}$. To each regular fractional ideal $\mathcal{E}$ of R we associate its *value semigroup ideal* $\Gamma_\mathcal{E} := \nu(\mathcal{E}^{\mathrm{reg}}) \subset \mathbb{Z}^{\mathfrak{V}_R}$. If $\mathcal{E} = R$, then the monoid Γ_R is called the *value semigroup* of R.

If $\mathcal{E}$ is a regular fractional ideal of R, then $\Gamma_\mathcal{E}$ is a semigroup satisfying particular properties, that we consider for any subset $E \subset \mathbb{Z}^s$:

(E0) there exists an $\alpha \in \mathbb{Z}^s$ such that $\alpha + \mathbb{N}^s \subset E$;

(E1) for any $\alpha, \beta \in E$, their component-wise minimum $\min\{\alpha, \beta\} \in E$;

(E2) for any $\alpha, \beta \in E$ with $\alpha_j = \beta_j$ for some j there exists an $\epsilon \in E$ such that $\epsilon_j > \alpha_j = \beta_j$ and $\epsilon_i \geq \min\{\alpha_i, \beta_i\}$ with equality if $\alpha_i \neq \beta_i$.

Definition 3 A submonoid S of $\mathbb{N}^s$ with group of differences $D_S = \mathbb{Z}^s$ is called a *good semigroup* if properties (E0), (E1), and (E2) hold for $E = S$.

A *semigroup ideal* of S is subset $E \subset \mathbb{Z}^s$ such that $E + S \subset E$ and $\alpha + E \subset S$ for some $\alpha \in \mathbb{Z}^s$. It is called a *good semigroup ideal* of the good semigroup S if it satisfies (E1) and (E2).

Proposition 4 *Let R be an admissible ring. Then,*

(i) the value semigroup Γ_R is a good semigroup;
(ii) for any regular fractional ideal $\mathcal{E}$ of R, $\Gamma_{\mathcal{E}}$ is a good semigroup ideal of Γ_R. □

On value semigroup ideals there is a distance function that mirrors the relative length of fractional ideals.

Definition 5 Let S be a good semigroup, and let $E \subset D_S$ be a subset. Then $\alpha, \beta \in E$ with $\alpha < \beta$ are called *consecutive* in E if $\alpha < \delta < \beta$ implies $\delta \notin E$ for any $\delta \in D_S$. For $\alpha, \beta \in E$, a chain of points $\alpha^{(i)} \in E$,

$$\alpha = \alpha^{(0)} < \cdots < \alpha^{(n)} = \beta, \tag{1}$$

is said to be *saturated of length n* if $\alpha^{(i)}$ and $\alpha^{(i+1)}$ are consecutive in E for all $i = 0, \ldots, n - 1$. If E satisfies

(E4) for fixed $\alpha, \beta \in E$, any two saturated chains (1) in E have the same length n;

then we call $d_E(\alpha, \beta) := n$ the *distance* of α and β in E.

D'Anna [4, Prop. 2.3] proved that any good semigroup ideal E satisfies property (E4).

Definition 6 For a good semigroup ideal E, the *conductor of E* is defined as $\gamma^E := \min\{\alpha \in E \mid \alpha + \mathbb{N}^s \subset E\}$. We denote $\gamma := \gamma^S$ and $\tau := \gamma - \mathbf{1}$.

Definition 7 Let S be a good semigroup, and let $E \subset F$ be two semigroup ideals of S satisfying property (E4). Then we call

$$d(F \backslash E) := d_F(\mu^F, \gamma^E) - d_E(\mu^E, \gamma^E)$$

the *distance* between E and F.

In the following, we collect the main properties of the distance function $d(-\backslash-)$. It follows from the definition that it is additive, as proven by D'Anna in [4, Prop. 2.7]:

Lemma 8 *Let $E \subset F \subset G$ be semigroup ideals of a good semigroup S satisfying properties (E1) and (E4). Then $d(G \backslash E) = d(G \backslash F) + d(F \backslash E)$.* □

Moreover, the distance function detects equality as formulated in [4, Prop. 2.8] and proved in [5, Prop. 4.2.6].

Proposition 9 *Let S be a good semigroup, and let E, F be good semigroup ideals of S with $E \subset F$. Then $E = F$ if and only if $d(F \backslash E) = 0$.* □

The length of a quotient of fractional ideals corresponds to the distance between the corresponding good semigroup ideals; see [4, Prop. 2.2] and [5, Prop. 4.2.7].

Proposition 10 *Let R be an admissible ring. If $\mathcal{E}, \mathcal{F}$ are two regular fractional ideals of R such that $\mathcal{E} \subset \mathcal{F}$ then, $\ell_R(\mathcal{F}/\mathcal{E}) = d(\Gamma_\mathcal{F} \backslash \Gamma_\mathcal{E})$.*   □

As a corollary, one can check equality of fractional ideals through their value semigroups:

Corollary 11 *Let R be an admissible ring, and let $\mathcal{E}, \mathcal{F}$ be two regular fractional ideals of R such that $\mathcal{E} \subset \mathcal{F}$. Then $\mathcal{E} = \mathcal{F}$ if and only if $\Gamma_\mathcal{E} = \Gamma_\mathcal{F}$.*   □

3   Canonical Ideals and Main Results

The following is the canonical semigroup ideal as defined by D'Anna in [4].

Definition 12 We call the semigroup ideal

$$K_S^0 := \left\{ \alpha \in \mathbb{Z}^s \mid \Delta^S(\tau - \alpha) = \emptyset \right\}.$$

the *normalized canonical semigroup ideal of S*, where

$$\Delta^S(\delta) := \Delta(\delta) \cap S = (\cup_{i \in I} \{\beta \in \mathbb{Z}^s \mid \delta i = \beta_i, \ \delta_j < \beta_j \ \forall \ j \neq i\}) \cap S$$

Definition 13 Let S be a good semigroup. Then S is called *symmetric* if $S = K_S^0$.

As mentioned in the introduction, Delgado proved that $S = \Gamma_R$ is symmetric if and only if R is Gorenstein. D'Anna [4] generalized this result: a regular fractional ideal $\mathcal{K}$ with $R \subset \mathcal{K} \subset \overline{R}$ is canonical if and only if $\Gamma_\mathcal{K} = K_S^0$. Recall that by definition a fractional ideal $\mathcal{K}$ is *canonical* if $\mathcal{K} : (\mathcal{K} : \mathcal{E}) = \mathcal{E}$ for any regular fractional ideal $\mathcal{E}$.

Definition 14 Let K be a good semigroup ideal of a good semigroup S. We call K a *canonical semigroup ideal* of S if $K \subset E$ implies $K = E$ for any good semigroup ideal E with $\gamma^K = \gamma^E$.

In analogy with this definition, we give a characterization of canonical semigroup ideals; see [5, Thm 5.2.7].

Theorem 15 *For a good semigroup ideal K of a good semigroup S the following are equivalent:*

(a) K is a canonical semigroup ideal;
(b) there exists an α such that $\alpha + K = K_S^0$;
(c) for all good semigroup ideals E one has $K - (K - E) = E$.

Moreover, if K satisfies these equivalent conditions, then $K - E$ is a good semigroup ideal for any good semigroup ideal E.   □

Given this characterization, it is natural to ask if taking the dual commutes with taking the semigroup. In the Gorenstein case, Pol [3] gave a positive answer.

Theorem 16 *If R is a Gorenstein admissible ring then,*

$$\Gamma_{R:\mathcal{E}} = \left\{\alpha \in \mathbb{Z}^s \mid \Delta^E(\tau - \alpha) = \emptyset\right\} = \Gamma_R - \Gamma_\mathcal{E}$$

for any regular fractional ideal $\mathcal{E}$ of R.

Our main result extends Pols result beyond the Gorenstein case.

Theorem 17 *Let $\mathcal{K}$ be a canonical ideal of R and let $K := \Gamma_\mathcal{K}$. Then, the following diagram commutes:*

$$
\begin{array}{ccc}
\left\{\begin{array}{c}\text{regular fractional}\\\text{ideals of } R\end{array}\right\} & \xrightarrow{\mathcal{E} \mapsto \mathcal{K}:\mathcal{E}} & \left\{\begin{array}{c}\text{regular fractional}\\\text{ideals of } R\end{array}\right\} \\
{\scriptstyle \mathcal{E} \mapsto \Gamma_\mathcal{E}}\Big\downarrow & & \Big\downarrow{\scriptstyle \mathcal{E} \mapsto \Gamma_\mathcal{E}} \\
\left\{\begin{array}{c}\text{good semigroup}\\\text{ideals of } \Gamma_R\end{array}\right\} & \xrightarrow{E \mapsto K - E} & \left\{\begin{array}{c}\text{good semigroup}\\\text{ideals of } \Gamma_R\end{array}\right\}
\end{array}
$$

Proof It is not restrictive to assume R local and $R \subset \mathcal{K} \subset \overline{R}$. Hence $K := \Gamma_\mathcal{K} = K_S^0$ by D'Anna [4].

Let $\mathcal{E} \subset \mathcal{F}$ be regular fractional ideals of R. Proposition 10 then yields

$$d(\Gamma_{\mathcal{K}:\mathcal{E}} \backslash \Gamma_{\mathcal{K}:\mathcal{F}}) = \ell_R((\mathcal{K} : \mathcal{E})/(\mathcal{K} : \mathcal{F})) = \ell_R(\mathcal{F}/\mathcal{E}) = d(\Gamma_\mathcal{F} \backslash \Gamma_\mathcal{E}) =: n.$$

Notice that $\ell_R((\mathcal{K} : \mathcal{E})/(\mathcal{K} : \mathcal{F})) = \ell_R(\mathcal{F}/\mathcal{E})$ as K is canonical. There is a composition series of regular fractional ideals

$$\mathcal{C}_\mathcal{E} = \mathcal{E}_0 \subsetneq \mathcal{E}_1 \subsetneq \cdots \subsetneq \mathcal{E}_l = \mathcal{E} \subsetneq \mathcal{E}_{l+1} \subsetneq \cdots \subsetneq \mathcal{E}_{l+n} = \mathcal{F},$$

where $\mathcal{C}_\mathcal{E}$ is the conductor of $\mathcal{E}$. By Corollary 11, applying Γ yields a chain of good semigroup ideals of Γ_R

$$C_{\Gamma_\mathcal{E}} = \Gamma_{\mathcal{E}_0} \subsetneq \Gamma_{\mathcal{E}_1} \subsetneq \cdots \subsetneq \Gamma_{\mathcal{E}_l} = \Gamma_\mathcal{E} \subsetneq \Gamma_{\mathcal{E}_{l+1}} \subsetneq \cdots \subsetneq \Gamma_{\mathcal{E}_{l+n}} = \Gamma_\mathcal{F}.$$

By Corollary 11 and Theorem 15(c), dualizing with K yields a chain of good semigroup ideals of Γ_R

$$\Gamma_{\mathcal{K}:\mathcal{C}_\mathcal{E}} = \Gamma_\mathcal{K} - \Gamma_{\mathcal{C}_\mathcal{E}} = K - C_{\Gamma_\mathcal{E}} = K - C_{\Gamma_{\mathcal{E}_0}} \supsetneq \cdots \supsetneq K - \Gamma_{\mathcal{E}_l} = K - \Gamma_\mathcal{E}$$
$$\supsetneq K - \Gamma_{\mathcal{E}_{l+1}} \supsetneq \cdots \supsetneq K - \Gamma_{\mathcal{E}_{l+n}} = K - \Gamma_\mathcal{F} \supset \Gamma_{\mathcal{K}:\mathcal{F}}. \quad (2)$$

By Theorem 15, $K - \Gamma_{\mathcal{E}_i}$ is a good semigroup ideal of S for all $i = 0, \dots, l + n$. Hence, using Proposition 9, we obtain $d(K - \Gamma_{\mathcal{E}_i} \backslash K - \Gamma_{\mathcal{E}_{i+1}}) \geq 1$ for all $i = 0, \dots, l + n - 1$. On the other hand, by Proposition 10,

$$d(\Gamma_{\mathcal{K}:\mathcal{C}_{\mathcal{E}}} \setminus \Gamma_{\mathcal{K}:\mathcal{F}}) = \ell_R(\mathcal{K} : \mathcal{C}_{\mathcal{E}}/\mathcal{K} : \mathcal{F}) = \ell_R(\mathcal{F}/\mathcal{C}_{\mathcal{E}}) = l + n.$$

By Lemma 8 and (2), it follows that $d(K - \Gamma_{\mathcal{E}_i} \setminus K - \Gamma_{\mathcal{E}_{i+1}}) = 1$ for all $i = 0, \ldots, l + n - 1$ and $d(K - \Gamma_{\mathcal{F}} \setminus \Gamma_{\mathcal{K}:\mathcal{F}}) = 0$. By Proposition 9 the latter is equivalent to the second claim.

In particular, this implies the following

Corollary 18 *Let $\mathcal{K}$ be a fractional ideal of an admissible ring R. Then $\mathcal{K}$ is canonical if and only if $K := \Gamma_{\mathcal{K}}$ canonical.*

References

1. E. Kunz, The value-semigroup of a one-dimensional Gorenstein ring. Proc. Am. Math. Soc. **25**, 748–751 (1970)
2. F. Delgado de la Mata, The semigroup of values of a curve singularity with several branches. Manuscr. Math. **59**(3), 347–374 (1987)
3. D. Pol, Logarithmic residues along plane curves, C.R. Math. Acad. Sci. Paris **353**(4), 345–349 (2015)
4. M. D'Anna, The canonical module of a one-dimensional reduced local ring. Commun. Algeb. **25**(9), 2939–2965 (1997)
5. P. Korell, M. Schulze, L. Tozzo, Duality of value semigroups. J. Commut. Algeb. Advance Publication, Rocky Mountain Mathematics Consortium (2018). https://projecteuclid.org:443/Euclid.jca/1473428763

Notes on Local Positivity and Newton–Okounkov Bodies

Harold Blum, Grzegorz Malara, Georg Merz and Justyna Szpond

Abstract We explore the notion of local numerical equivalence in higher dimension and its relationship with Newton–Okounkov bodies with respect to flags centered at a given point.

1 Introduction

Let D be a big divisor on a smooth projective variety X of dimension d. The *Newton–Okounkov body* of D serves as a tool for studying positivity properties of D. The construction of the Newton–Okounkov body was first introduced in the work of Okounkov [1] and independently developed in the work Lazarsfeld–Mustață [2] and Kaveh–Khovanskii [3]. The construction is dependent on a flag

$$Y_\bullet = \{X = Y_0 \supset Y_1 \supset \cdots \supset Y_d = \{p\}\}$$

We thank Francesco Bastianelli, Magdalena Lampa-Baczyńska, Tomasz Szemberg and Halszka Tutaj-Gasińska for fruitful conversations. HB received support from NSF, grant DMS-0943832. JS was partially supported by National Science Centre, Poland, grant 2014/15/B/ST1/02197. GeM received support from DFG Research Training Group 1493

H. Blum (✉)
Department of Mathematics, University of Michigan, Ann Arbor, MI 48105, USA
e-mail: blum@umich.edu

G. Malara · J. Szpond
Pedagogical University of Cracow, Department of Mathematics, Podchorążych 2,
30-084 Kraków, Poland
e-mail: grzegorzmalara@gmail.com

J. Szpond
e-mail: szpond@gmail.com

G. Merz
Mathematisches Institut, Georg-August Universität Göttingen Bunsenstraße 3–5,
37073 Göttingen, Germany
e-mail: georg.merz@mathematik.uni-goettingen.de

M. Alberich-Carramiñana et al. (eds.), *Extended Abstracts February 2016*,
Trends in Mathematics 9, https://doi.org/10.1007/978-3-030-00027-1_17

such that each Y_i is smooth at p. The *Newton–Okounkov body of D along $Y_\bullet$* is a convex set $\Delta_{Y_\bullet}(D) \subset \mathbb{R}^d$ and encodes information on the vanishing of sections of $H^0(\mathcal{O}_X(mD))$ along $Y_\bullet$. The Newton–Okounkov bodies of a divisor D are only dependent on the numerical equivalence class of D. The following theorem states the relationship between Newton–Okounkov bodies and numerical equivalence.

Theorem 1 ([2, 4]) *Let D_1 and D_2 be big divisors on a smooth projective variety X. The following are equivalent:*

(a) for all admissible flags $Y_\bullet$ on X, we have $\Delta_{Y_\bullet}(D_1) = \Delta_{Y_\bullet}(D_2)$;
(b) the divisors D_1 and D_2 are numerically equivalent.

Philosophically, Theorem 1 implies that all numerical properties of a divisor D are encoded in the convex geometry of Newton–Okounkov bodies of D. For example, the volume of a divisor D is $d!$ times the Euclidean volume of $\Delta_{Y_\bullet}(D) \subset \mathbb{R}^d$.

In [5–7], it was shown that local positivity at some point $O \in X$ is related to the Newton–Okounkov bodies of D with respect to admissible flags centered at the point O. Motivated by these ideas, Roé asks the following.

Question 2 *Let D_1 and D_2 be big divisors on X such that $\Delta_{Y_\bullet}(D_1) = \Delta_{Y_\bullet}(D_2)$ for all admissible flags $Y_\bullet$ centered at O. How are D_1 and D_2 related?*

When X is a surface, Roé [8] gives an elegant answer to this question. First, he introduces the following definition.

Definition 3 Let D be a big divisor on a smooth projective surface X and write $D = P(D) + N(D)$ for the Zariski decomposition of D into positive and negative components. Next, fix a point $O \in X$ and write $N(D) = N_O(D) + N_O^c(D)$ for the decomposition of $N(D)$ into components containing O and disjoint from O. We say that two big divisors D_1 and D_2 are *locally numerically equivalent at O* if $P(D_1) \equiv P(D_2)$ and $N_O(D_1) = N_O(D_2)$. Note that local numerical equivalence at all points of X implies numerical equivalence.

Roughly speaking, two divisors are locally numerically equivalent at a point O if the divisors are numerically equivalent modulo fixed components of D that do not pass through O. With this definition, Roé proves the following.

Theorem 4 (Roé, [8]) *Let D_1 and D_2 be big divisors on a smooth projective surface X and $O \in X$ a closed point. The following are equivalent:*

(a) for all admissible flags $Y_\bullet$ centered over O, we have $\Delta_{Y_\bullet}(D_1) = \Delta_{Y_\bullet}(D_2)$;
(b) the divisors D_1 and D_2 are locally numerically equivalent at O.

Roé leaves the generalization of Theorem 4 to higher dimensions open. A key obstacle in extending the theorem to higher dimensions is that Zariski decompositions of big divisors do not always exist in dimensions three and higher. However, Nakayama [9] introduced a weaker analogue of the Zariski decomposition called the σ-decomposition. Such decompositions always exist for big divisors. (See Sect. 4 for the definition of the σ-decomposition.)

We extend Roé's definition of local numerical equivalence to higher dimensions by replacing the Zariski decomposition in the definition with the σ-decomposition.

Definition 5 Let D be a big divisor on a smooth projective variety X and write $D = P_\sigma(D) + N_\sigma(D)$ for the σ-decomposition of D. Next, fix a point $O \in X$ and write $N_\sigma(D) = N_{\sigma,O}(D) + N_{\sigma,O}^c(D)$ for the decomposition of $N_\sigma(D)$ into components containing O and disjoint from O. We say that two big divisors D_1 and D_2 on X are *locally numerically equivalent at* O if $P_\sigma(D_1) \equiv P_\sigma(D_2)$ and $N_{\sigma,O}(D_1) = N_{\sigma,O}(D_2)$.

Using this definition, we conjecture the following generalization of Theorem 4 to higher dimensions. (We plan to prove this conjecture in a forthcoming paper. An idea of the proof will be given at the end of Sect. 4.)

Conjecture 6 *Let D_1, and D_2 be two big divisors on a smooth projective variety X and $O \in X$ a closed point. The following are equivalent:*

(a) for all admissible flags $Y_\bullet$ centered over O, we have $\Delta_{Y_\bullet}(D_1) = \Delta_{Y_\bullet}(D_2)$;
(b) the divisors D_1 and D_2 are locally numerically equivalent at O.

2 Preliminaries

For the purposes of this paper all varieties will be defined over the complex numbers. A divisor D on a variety X will always mean a Cartier divisor. See Lazarsfeld [10] for basic properties of divisors and linear series.

2.1 Newton–Okounkov Bodies

Given a big divisor D on X, we seek to understand the sections of $H^0(X, \mathcal{O}_X(mD))$ as m grows. The construction of Newton–Okounkov bodies will encode such sections in the form of a convex body. Before explaining the construction, we define the following.

Definition 7 Let X be a normal projective variety of dimension d. We call $Y_\bullet$ an *admissible flag on* X if $Y_\bullet$ is a flag on X, where

$$Y_\bullet = \{X = Y_0 \supset Y_1 \supset \cdots \supset Y_d = \{p\}\}$$

such that each Y_i is an irreducible subvariety of codimension i and is smooth at the point p. We say that $Y_\bullet$ is an *admissible flag over* X if there is a proper birational morphism $\pi : \tilde{X} \to X$ such that $Y_\bullet$ is an admissible flag on $\tilde{X}$. (Note that there is a distinction between an admissible flag *on* X and an admissible flag *over* X.)

Fix a point $O \in X$. A flag $Y_\bullet$ is an *admissible flag over* O if $Y_\bullet$ is an admissible flag over X and π (as above) contracts Y_d to O.

We now proceed to define the *Newton–Okounkov body* associated to a big divisor D on X and an admissible flag $Y_\bullet$ on X. Given a divisor F on X, there is a valuation map

$$\nu_{Y_\bullet} = \nu \colon H^0(X, \mathcal{O}_X(F)) \setminus \{0\} \longrightarrow \mathbb{Z}^d_{\geq 0}$$

that measures order of vanishing of sections along $Y_\bullet$; see Lazarsfeld–Mustaţă [2] for the precise definition of $\nu_{Y_\bullet}$. We write $\nu_{Y_\bullet}(F)$ for the image of the map. If $Y_\bullet$ is an admissible flag on X, then the *Newton–Okounkov body of D along $Y_\bullet$* is the convex body

$$\Delta_{Y_\bullet}(D) := \text{closed convex hull} \left(\bigcup_{m \geq 1} \frac{1}{m} \nu_{Y_\bullet}(mD) \right).$$

If $Y_\bullet$ is a flag over X (not necessarily on X) such that $\pi \colon \tilde{X} \to X$ is a proper birational morphism and $Y_\bullet$ is a flag on $\tilde{X}$, then the *Newton–Okounkov body of D along $Y_\bullet$* is $\Delta_{Y_\bullet}(D) := \Delta_{Y_\bullet}(\pi^*D)$, where the latter Newton–Okounkov body is computed on $\tilde{X}$.

2.2 Decomposition of Big Divisors

Now, we recall some information on decompositions of big divisors. Let X be a smooth projective variety of dimension d, and D a big divisor on X. A decomposition $D = P + N$ is said to be a *Zariski decomposition* if P and N are $\mathbb{R}$-divisors on X such that: (i) P is nef; (ii) N is effective; and (iii) the natural map

$$H^0(X, \mathcal{O}_X(\lfloor mP \rfloor)) \longrightarrow H^0(X, \mathcal{O}_X(mD))$$

is an isomorphism for all $m \in \mathbb{Z}_{>0}$. When X is a surface, such decompositions of big divisors always exist by the work of Fujita. In higher dimensions, such decompositions do not always exist.

However, Nakayama introduced a weaker notion known as the σ-decomposition [9]. The σ-decomposition of a big divisor always exists and is defined as follows. Let D be a big divisor on X and Γ a prime divisor on X. We first define

$$\sigma_\Gamma(D) := \lim_{m \to \infty} \frac{\text{ord}_\Gamma |mD|}{m}.$$

Note that if $|F|$ is a linear system and Γ is a prime divisor, then $\text{ord}_\Gamma |F|$ denotes the coefficient of Γ in a general element of $|F|$.

The σ-decomposition of X is given by $D = P_\sigma(D) + N_\sigma(D)$, where

$$N_\sigma(D) := \sum_\Gamma \sigma_\Gamma(D)\Gamma$$

with the previous sum running over all prime divisors Γ on X. Thus, $P_\sigma(D) := D - N_\sigma(D)$. Note that $N_\sigma(D)$ and $P_\sigma(D)$ are $\mathbb{R}$-divisors, but not necessarily $\mathbb{Q}$-divisors.

It is an easy exercise to see that the natural map

$$H^0(X, \mathcal{O}_X(mD)) \longrightarrow H^0(X, \mathcal{O}_X(\lfloor mP \rfloor))$$

is an isomorphism. Thus, $D = P_\sigma(D) + N_\sigma(D)$ is a Zariski decomposition if P is nef.

Example 8 Let $\pi: B_O(\mathbb{P}^2) \to \mathbb{P}^2$ be the blow up of $\mathbb{P}^2$ at a point O and E be the exceptional divisor of π. Let L be the inverse image of a line in $\mathbb{P}^2$ not passing through O. We proceed to consider different Newton–Okounkov bodies for the big divisor $D := L + E$. Note that D has Zariski decomposition given by $P(D) = L$ and $N(D) = E$. First, we consider the flag $Y_\bullet = \{B_O(\mathbb{P}^2) \supset C \supset \{p\}\}$, where C is the inverse image of a line in $\mathbb{P}^2$ not containing O and p is an arbitrary point in C. The Newton–Okounkov body $\Delta_{Y_\bullet}(D)$ is the following polytope:

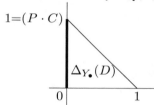

Note that this Newton–Okounkov body allows us to compute $(P \cdot C)$ as the volume of the segment of $\Delta_{Y_\bullet}(D)$ with the y-axis.

Now, we consider the flag $W_\bullet = \{B_O(\mathbb{P}^2) \supset E \supset \{Q\}\}$, where Q is a point on the exceptional divisor E. The Newton–Okounkov body $\Delta_{W_\bullet}(D)$ is given by:

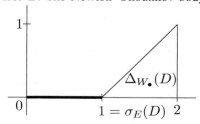

This Newton–Okounkov body allows us to compute the coefficient $\sigma_E(D)$ of E in N which is given by the minimal coordinate of the projection of $\Delta_{W_\bullet}(D)$ to the x-axis.

3   Newton–Okounkov Bodies and Numerical Equivalence

In this section, we seek to explain the ideas behind the proof of Theorem 1. The statement (b) implies (a) was proved by Lazarsfeld–Mustață [2, Prop. 4.1]. Their argument relies on Fujita's Vanishing Theorem and is brief. The reverse implication was proved by Jow [4, Thm. A] and is more involved.

The key idea in the proof of [4, Thm. A] is the following. For a big divisor D on X and a curve $C \subset X$, we would like to read $D \cdot C$ off from a Newton–Okounkov body $\Delta_{Y_\bullet}(D)$, where $Y_{d-1} = C$. Jow shows this is possible for very general complete intersection curves C and specially chosen flags $Y_\bullet$.

We explain the result of [4] in further detail. Let $A_1, \ldots, A_{d-1}$ be very ample divisors on a projective variety X such that $Y_\bullet$ is an admissible flag with

$$Y_r = A_1 \cap \cdots \cap A_r$$

for $i = 1, \ldots, d - 1$. If $A_1, \ldots, A_{d-1}$ are chosen very generally, [4, Cor. 3.3 & Thm. 3.4] relates $D \cdot Y_{d-1}$ to the Euclidean volume of $\Delta_{Y_\bullet}(D)$ restricted to the subspace of $\mathbb{R}^d$ whose first $d - 1$ coordinates are 0.

We proceed to give the precise relationship. For a very general choice of $A_1, \ldots, A_{d-1}$,

$$\mathrm{vol}(\Delta_{Y_\bullet}(D)|_{\mathbf{0}^{d-1}}) = Y_{d-1} \cdot D - \sum_{i=1}^{n} \sum_{p \in Y_{d-1} \cap E_i} \sigma_{E_i}(D),$$

where $E_1, \ldots, E_n$ are the irreducible divisorial components of the base locus of $|D|$. In [4], the author requires $A_1, \ldots, A_{d-1}$ to be chosen so that Y_{d-1} has well behaved intersection with the base locus of $|D|$. Additionally, the requirements on $A_1, \ldots, A_{d-1}$ imply that Y_d does not lie in the base locus of $|D|$.

4   Newton–Okounkov Bodies and Local Numerical Equivalence

In this section we seek to explain the extension of Theorem 4 to higher dimensions. We believe that the natural extension of Definition 3 to higher dimensions is achieved by replacing Zariski decompositions with σ-decompositions (see Definition 5).

Note that in dimension two, our definition of local numerical equivalence agrees with Roé's definition (Definition 3). Indeed, for surfaces the σ-decomposition agrees with the standard Zariski decomposition [9, Rem. III.1.17]. With our definition, we conjecture the following.

Conjecture 9 *Let D_1 and D_2 be two big divisors on a smooth projective variety X, and $O \in X$ a closed point. The following are equivalent:*

(a) for all admissible flags $Y_\bullet$ centered over O, $\Delta_{Y_\bullet}(D_1) = \Delta_{Y_\bullet}(D_2)$;
(b) the divisors D_1 and D_2 are locally numerically equivalent at O.

We now present the main ideas how to prove the conjecture. The implication (b) implies (a) follows easily from the fact that the Newton–Okounkov body of a big divisor D is only dependent on the the numerical equivalence class of D and the equality $\Delta_{Y_\bullet}(D) = \Delta_{Y_\bullet}(P_\sigma(D) + N_{\sigma,O}(D))$ for all big divisors D on X.

Indeed, the above equality can be derived by showing that the bijection of sections

$$\Phi \colon H^0(X, \mathcal{O}_X(mD)) \xrightarrow{\sim} H^0(X, \mathcal{O}_X(\lfloor P_\sigma(mD) + N_{\sigma,O}(mD)\rfloor))$$

is compatible with the valuation map $\nu_{Y_\bullet}$. It is also worth to note that this equality can be easily deduced from Küronya–Lozovanu [7, Thm. 4.2(3)], which says that $\Delta_{Y_\bullet}(D) = \Delta_{Y_\bullet}(P_\sigma(D)) + \nu_{Y_\bullet}(N_\sigma(D))$.

The implication (a) implies (b) is more involved. The main idea is to use the same technique as in the proof [4], i.e., constructing flags $Y_\bullet$ that encode numerical data of a fixed big divisor. However, we have the additional constraint that the flag $Y_\bullet$ must be centered at the point O. The techniques of Jow extend to our setting if the base locus of $|mD_i|$ contains no embedded components at O and the reduced base locus of $|mD_i|$ is smooth at O for all m. If this is not the case, then we need to consider the blow up $B_O X \to X$ at O with exceptional divisor E. For a very general $O' \in E$, the previous conditions will be satisfied for O replaced with O' and D replaced with $\pi^*(D)$. Thus, it is essential that we consider flags over O (which are flags on birational models over X), instead of simply flags on X.

References

1. A. Okounkov, Brunn-Minkowski inequality for multiplicities. Invent. Math. **125**, 405–411 (1996)
2. R. Lazarsfeld, M. Mustaţă, Convex bodies associated to linear series. Ann. Sci. Éc. Norm. Supr. (4) **42**(5), 783–835 (2009)
3. K. Kaveh, A.G. Khovanskii, Newton–Okounkov bodies, semigroups of integral points, graded algebras and intersection theory. Ann. of Math. (2) **176**(2), 925–978 (2012)
4. S.-Y. Jow, Okounkov bodies and restricted volumes along very general curves. Adv. Math. **223**, 1356–1371 (2010)
5. A. Küronya, V. Lozovanu, Local positivity of linear series on surfaces (2014). arXiv:1411.6205. preprint
6. A. Küronya, V. Lozovanu, Infinitesimal Newton–Okounkov bodies and jet separation (2015). arXiv:1507.04339. preprint
7. A. Küronya, V. Lozovanu, Positivity of line bundles and Newton–Okounkov bodies (2015). arXiv:1506.06525. preprint
8. J. Roé, Local positivity in terms of Newton–Okounkov bodies (2016). arXiv:1505.02051. preprint
9. N. Nakayama, *Zariski decomposition and abundance* (Mathematical Society of Japan, Tokyo, 2004)
10. R. Lazarsfeld, *Positivity in Algebraic Geometry I* (Springer, Berlin, 2004)

Newton–Okounkov Bodies and Reified Valuations of Higher Rank

Alberto Camara, Iago Giné, Roberto Gualdi, Nikita Kalinin, Joaquim Roé, Martin Ulirsch, Stefano Urbinati and Xavier Xarles

Abstract We study the shape change of the Newton–Okounkov body of a fixed divisor D with respect to a valuation v moving in a suitable space of (higher-rank) valuations.

A. Camara
School of Mathematical Sciences, University of Nottingham, University Park,
Nottingham NG7 2RD, Nottingham, UK
e-mail: alberto.camara@nottingham.ac.uk

I. Giné · J. Roé (✉) · X. Xarles
Departament de Matemàtiques, Universitat Autònoma de Barcelona,
08193 Barcelona, Bellaterra, Spain
e-mail: jroe@mat.uab.cat

I. Giné
e-mail: iagogv@mat.uab.cat

X. Xarles
e-mail: xarles@mat.uab.cat

R. Gualdi
Institut de Mathématiques de Bordeaux, Université de Bordeaux,
33405 Talence, France
e-mail: roberto.gualdi@math.u-bordeaux.fr

N. Kalinin
University of Geneva, Section de Mathematiques, rte de Drize 7,
villa Battelle, 1227 Geneva, Switzerland
e-mail: nikaanspb@gmail.com

M. Ulirsch
Hausdorff Center for Mathematics, University of Bonn, 53115 Bonn, Germany
e-mail: ulirsch@math.uni-bonn.de

S. Urbinati
Dipartimento di Matematica, Università degli Studi di Padova,
35121 Padova, Italy
e-mail: urbinati.st@gmail.com

© Springer Nature Switzerland AG 2018
M. Alberich-Carramiñana et al. (eds.), *Extended Abstracts February 2016*,
Trends in Mathematics 9, https://doi.org/10.1007/978-3-030-00027-1_18

1 Introduction

Newton–Okounkov bodies are a modern embodiment of a classical technique in algebraic geometry, namely to associate a simple polyhedral object to an algebraic variety X (possibly with additional data) and recover deep geometric properties from the geometry of the polyhedron. The very first example here is the Newton polygon associated to a plane algebraic curve that gives information about the genus.

The definition of Newton–Okounkov bodies originates in papers due to A. Okounkov from the middle of the 1990s as a generalization of both the Newton polygon for toric hypersurfaces and the moment polytope for toric varieties. More recently, Lazarsfeld–Mustață [1] and, independently, Kaveh–Khovanskii [2] defined generally the Newton–Okounkov body $\Delta_v(D)$ on X for a divisor D and a valuation v of maximal rank on the function field of X.

The shape of Newton–Okounkov bodies has been studied from several points of view, notably with regard to its connection with local positivity. In fact, in several different cases, it is possible to recover Seshadri-type invariants associated to the linear series just looking at the convex body, see for example [3–5]. In most previous works, with [3] being a partial exception, the valuation v is fixed throughout the paper.

The goal of our group project was to understand how the shape of the body changes while, fixing the divisor D, the valuation v is moving in a suitable space of (higher-rank) valuations.

2 Spaces of Reified Higher Rank Valuations

Throughout this report we denote by $\mathbb{R}^{(k)}$ (and similarly by $\mathbb{Z}^{(k)}$) the additive group $(\mathbb{R}^k, +)$ (respectively $(\mathbb{Z}^k, +)$), endowed with the lexicographic order $\leq = \leq_{\text{lex}}$. That is, for two vectors $x = (x_1, \ldots, x_k)$ and $(y_1, \ldots, y_k)$ we have $x < y$ if and only if there is $1 \leq r \leq k$ such that $x_i = y_i$ for all $i < r$ and $x_r < y_r$.

Definition 1 Let K be a field. A *(reified) valuation* on K is a function

$$\text{val}: K^* \longrightarrow \mathbb{R}^{(k)}$$

satisfying $\text{val}(ab) = \text{val}(a) + \text{val}(b)$ as well as $\text{val}(a + b) \geq \min\{\text{val}(a), \text{val}(b)\}$ for all $a, b \in K^*$. We say that a valuation is *discrete*, if it factors as

$$\text{val}: K^* \to \mathbb{Z}^{(k)} i \mathbb{R}^{(k)},$$

where i is an order-preserving homomorphism.

Note hereby that any order-preserving monomorphism $i: \mathbb{Z}^{(n)} \hookrightarrow \mathbb{R}^{(n)}$ is given by $(l_1, \ldots, l_n) \longmapsto (a_1 l_1, \ldots, a_n l_n)$, where $a_i \in \mathbb{R}_{>0}$ for all $i = 1, \ldots, n$. When we want to stress the dependence of i on the a_i we also write $i = i_{\mathbf{a}}$ for $\mathbf{a} = (a_1, \ldots, a_n)$. As usual, we may extend every valuation to K by setting $\mathrm{val}(0) = \infty$. We denote by $\overline{\mathbb{R}}^{(k)}$ (or $\overline{\mathbb{Z}}^{(k)}$) the extended ordered monoids $(\mathbb{R}^k \sqcup \{\infty\}, +, \leq_{\mathrm{lex}})$ (or $(\mathbb{Z}^k \sqcup \{\infty\}, +, \leq_{\mathrm{lex}})$ respectively).

Recall that a subgroup Γ' of an ordered group Γ is said to be *convex*, if every $\gamma \in \Gamma$ that fulfills $\gamma_1' \leq \gamma \leq \gamma_2'$ for some $\gamma_i' \in \Gamma'$ lies already in Γ'. The *rank of a finitely generated ordered group* Γ is the maximal length of a flag

$$\Gamma = \Gamma_0 \supsetneq \Gamma_1 \supsetneq \cdots \supsetneq \Gamma_k = \{0\}$$

of convex subgroups. Using this we may define the *rank of a valuation* as the rank of $\mathrm{val}(K^*)$.

Example 2 Let X be a normal complex projective variety of dimension n. Consider a flag of irreducible subvarieties

$$Y_\bullet : X = Y_0 \supsetneq Y_1 \supsetneq \ldots \supsetneq Y_{n-1} \supsetneq Y_n = \{pt\}$$

such that $\mathrm{codim}\, Y_i = i$ and Y_n is a smooth point on each Y_i. For every non-zero rational function $s \in K(X)$, set $s_0 := s$, and inductively define for $i = 1, \ldots, n$

$$\nu_i(s) := \mathrm{ord}_{Y_i}(s_{i-1})$$

$$s_i := \left. \frac{s_{i-1}}{g_i^{\nu_i(s)}} \right|_{Y_i},$$

where g_i is the local equation of Y_i in Y_{i-1} near Y_n. The association

$$\mathrm{val}_{Y_\bullet} : K(X) \longrightarrow \overline{\mathbb{Z}}^{(n)}$$
$$s \longmapsto (\nu_1(s), \ldots, \nu_n(s))$$

defines a discrete valuation with value group $\mathbb{Z}^{(n)}$ of rank n on the function field of X. For every $\mathbf{a} = (a_1, \ldots, a_n) \in \mathbb{R}^n_{>0}$, we therefore obtain a reified valuation $v_{Y_\bullet, \mathbf{a}} = i_{\mathbf{a}} \circ v_{Y_\bullet}$.

Given a field K, we write Val_K^k for the set of all (reified) valuations $\mathrm{val} \colon K^* \to \mathbb{R}^{(k)}$. The set Val_K carries a natural topology. It is given as the coarsest topology making the evaluation functions

$$\mathrm{ev}_f \colon \mathrm{Val}_K^k \longrightarrow \mathbb{R}^{(k)}$$
$$\mathrm{val} \longmapsto \mathrm{val}(f)$$

for all $f \in K^*$ continuous, where the topology on $\mathbb{R}^{(k)}$ is the Euclidean one.

Note that the (order-preserving) projections $\mathrm{pr}_{k,l} \colon \mathbb{R}^{(k)} \to \mathbb{R}^{(l)}$ onto the first l coordinates for $l \leq k$ induce natural continuous maps $\mathrm{pr}_{k,l} \colon \mathrm{Val}_K^k \to \mathrm{Val}_K^l$. For a variety X, we write $\mathrm{Val}_X^k = \mathrm{Val}_{K(X)}^k$.

Remark 3 The space Val_X^k is the locus of birationally invariant points X_k^{bir} in the *Hahn analytification* $X_k^{\mathfrak{H}}$ of X, as introduced in [6, 7]. In particular, for $k = 1$ we obtain the set of birationally invariant points X^{bir} in the Berkovich analytification X^{an} of X; see [8]. Note that for technical (model-theoretic) reasons the authors of [6] also choose to endow $\mathbb{R}^{(k)}$ (and thereby also Val_X^k) with the lexicographic topology, i.e., the topology generated by lexicographic intervals, and not with the Euclidean one.

3 Newton–Okounkov Bodies

Let X be normal complex projective variety of dimension n and consider a divisor D on X. For a valuation $v \in \mathrm{Val}_X^n$ the semigroup of valuation vectors is defined by

$$\Gamma_v(D) := \{(v(s), m) \in \mathbb{R}^n \times \mathbb{N} | s \in H^0(X, \mathcal{O}(mD))\} \subseteq \mathbb{R}^n \times \mathbb{N} .$$

Definition 4 The *Newton–Okounkov body* associated to D and v is given by

$$\Delta_v(D) := \mathrm{cone}_{\mathbb{R}^n \times \mathbb{R}_{\geq 0}} (\Gamma_v(D)) \cap (\mathbb{R}^n \times \{1\}) .$$

The above Definition 4 is generalization of the one given in [2] which corresponds to the case of the valuation v being discrete with value group $\mathbb{Z}^{(n)} \subseteq \mathbb{R}^{(n)}$; it differs from the notion considered in Boucksom [9], in that we do not take the rational rank of v into account. In the special case that $v = v_{Y_\bullet}$, as in Example 2, we obtain the construction that is studied in Lazarsfeld–Mustață [1].

Example 5 Let X be a smooth projective surface. In this case the construction of the Newton–Okounkov body is strictly connected to the well known fact that Zariski decomposition exists for surfaces, i.e., any pseudoeffective divisor D can be written as $D = P_D + N_D$, where P_D is nef, we have $P_D.N_D = 0$, and N_D is effective with a negatively defined intersection matrix.

Let us consider the rank two valuation induced by a general flag $Y_\bullet = \{X \supseteq C \supseteq p\}$ such that $p \notin \mathrm{supp}(N_D)$. In Lazarsfeld–Mustață [1] the authors give two which compute the boundary of the Newton–Okounkov body.

Let $\alpha(D) = \mathrm{ord}_p(N_D)$ and $\beta(D) = \alpha(D) + C.P_D$ then, given $\mu := \sup\{t | D - tC$ is big$\}$ we have $\Delta_{Y_\bullet}(D) = \{(x, y) \in \mathbb{R}^2 | 0 \leq x \leq \mu$ and $\alpha(D - xC) \leq y \leq \beta(D - xC)\}$.

The goal of our project is to study the dependence of $\Delta_v(D)$ on the valuation $v \in \mathrm{Val}_X^n$. Our first observation is the following:

Conjecture 6 (Continuity Principle) *Let $F(\mathbb{R}^n)$ be the set of non-empty compact subsets of $\mathbb{R}^n$, endowed with the Hausdorff distance. Then, the association*

$$\Delta(D)\colon Val_X^n \longrightarrow F(\mathbb{R}^n)$$
$$v \longmapsto \Delta_v(D)$$

is continuous.

So far, in Ciliberto–Farnik–Küronya–Lozovanu–Roé–Shramov [3] the dependence of $\Delta_v(D)$ has only been investigated from the point of view of $\mathrm{pr}_{n,1}(v)$ varying in Val_X^1. The following reinterpretation of the example considered in [3, § 5.4.2] gives us a hint why this may not be the necessary generality.

Example 7 Let $X = \mathbb{P}^2$ and C a germ of a curve at the origin 0 of $\mathbb{A}^2 = \mathrm{Spec}\mathbb{C}[x, y]$ $\subseteq \mathbb{P}^2$ that is tangent to the line $y = 0$. Locally around 0 we may parametrize C by $x \mapsto (x, \xi(x)) \in \mathbb{A}^2$ where $\xi(x) \in \mathbb{C}[[x]]$ with $\xi(0) = \xi'(0) = 0$. In [3, Def. 3.1 and Prop. 3.10] the authors consider a family $v(C, s; .) = \big(v_1(C, s; .), v_2(C, s; .)\big)$ of valuations in Val_X^2 parametrized by $s \in (1, \infty)$.

For $f \in \mathbb{C}(x, y)$, write f as a Laurent series $f = \sum a_{ij} x^i \big(y - \xi(x)\big)^j$. In these coordinates the first component of $v(C, s; .)$ is given by

$$v_1(C, s; f) = \min\big\{i + sj \big| a_{ij} \neq 0\big\}$$

and the second component by

$$v_2(C, s; f) = \min\big\{j \big| \exists i \text{ such that } a_{ij} \neq 0 \text{ and } i + sj = v_1(C, s; f)\big\}.$$

Suppose now that C is a conic. By [3, Proposition 5.24] we have

$$\Delta_{v(C,s,.)} = \begin{cases} \Delta_{1,s,1} & \text{if } 1 < s < 2, \\ \Delta_{2,\frac{s}{2},\frac{1}{2}} & \text{if } s \geq 2 \end{cases}$$

where $\Delta_{a,b,c}$ denotes the triangle in $\mathbb{R}^2$ whose vertices are given by $(0, 0)$, $(a, 0)$, (b, c).

As the association $s \mapsto v_1(C, s; .)$ defines a continuous map $(1, \infty) \to Val_X^1$ at first sight this appears to be a counterexample to Conjecture 6 above. But it is not, since $s \mapsto v_2(C, s; .)$ is not continuous and, in general, we need both components to be continuous to obtain a continuous path in Val_X^2.

4   Volumes of Newton–Okounkov Bodies

Definition 8 Let $v\colon K^* \to \mathbb{R}^{(n)}$ be a valuation on a field K with valuation ring R and consider the ideals $\mathfrak{q}_m = \big\{q \in R \big| |v(m)|_1 \geq m\big\}$, for $m \in \mathbb{Z}$, where we write $|\mathbf{r}|_1 = |r_1| + \cdots + |r_n|$ for $\mathbf{r} = (r_1, \ldots, r_n) \in \mathbb{R}^n$. The *volume of v* is defined to be

$$\text{vol}(v) = \limsup_{m \to \infty} \frac{\text{length}(R/\mathfrak{q}_m)}{m^n/n!} \, .$$

Let X be a normal complex projective variety of dimension n and D a divisor on X.

Conjecture 9 *For every valuation v on $K(X)$ of rank n, $\text{vol}(\Delta_v(D)) = \frac{1}{n!}\text{vol}(D)$ $\text{vol}(v)$.*

Note that in the special case that v is a discrete valuation with image $\mathbb{Z}^{(n)} \subseteq \mathbb{R}^{(n)}$ we have $\text{vol}(v) = 1$ and therefore the above formula reduces to

$$\text{vol}(\Delta_v(D)) = \frac{1}{n!}\text{vol}(D) \, ,$$

which is well-known; see Lazarsfeld–Mustață [1, Thm. 2.3]. Let us now outline a strategy, with which to approach Conjecture 9:

(i)   If $v = v_{Y_\bullet, \mathbf{a}}$ for a flag $Y_\bullet$ and $\mathbf{a}$ as in Example 2, we can directly verify the claim.
(ii) More generally, by [3], every discrete valuation is given by a flag $Y_\bullet$ on a suitable birational model $X' \to X$ of X. Therefore we expect that the same argument will work for all discrete valuations.
(iii) In order to establish the general case of the conjecture we first need to show that discrete valuations are dense in $\text{Val}_{K(X)}$. Then the claim would follow from both $v \mapsto \text{vol}(\Delta_v(D))$ and $v \mapsto \text{vol}(v)$ depending continuously on v.

References

1. R. Lazarsfeld, M. Mustață, Convex bodies associated to linear series, Ann. Sci. Éc. Norm. Supér. (4) **42**(5), 783–835 (2009)
2. K. Kaveh, A.G. Khovanskii, Newton–Okounkov bodies, semigroups of integral points, graded algebras and intersection theory, Ann. of Math. (2) **176**(2), 925–978 (2012)
3. C. Ciliberto, M. Farnik, A. Küronya, V. Lozovanu, J. Roé, C. Shramov, Newton–Okounkov bodies sprouting on the valuative tree (2016). arXiv:1602.02074
4. A. Küronya, V. Lozovanu, Local positivity of linear series (2014). arXiv:1411.6205
5. A. Küronya, V. Lozovanu, Infinitesimal Newton–Okounkov bodies and jet separation (2015). arXiv:1507.04339
6. T. Foster, D. Ranganathan, Hahn analytification and connectivity of higher rank tropical varieties (2015). arXiv:1504.07207
7. K.S. Kedlaya, Reified valuations and adic spectra (2013). arXiv:1309.0574
8. V. Berkovich, *Spectral Theory and Analytic Geometry Over Non-archimedean Fields*, vol. 33, Mathematical surveys and monographs (American Mathematical Society, Providence, 1990)
9. S. Boucksom, *Corps d'Okounkov [d'après Okounkov, Lazarsfeld–Mustață et Kaveh–Khovanskii]*, Séminaire Bourbaki 1059 (2012), pp. 38